RAND NATIONAL DEFENSE RESEARCH INSTITUTE

DoD Depot-Level Reparable Supply Chain Management

Process Effectiveness and Opportunities for Improvement

Eric Peltz, Marygail K. Brauner, Edward G. Keating, Evan Saltzman, Daniel Tremblay, Patricia Boren

Prepared for the Office of the Secretary of Defense

The research described in this report was prepared for the Office of the Secretary of Defense (OSD). The research was conducted within the RAND National Defense Research Institute, a federally funded research and development center sponsored by OSD, the Joint Staff, the Unified Combatant Commands, the Navy, the Marine Corps, the defense agencies, and the defense Intelligence Community under Contract W91WAW-12-C-0030.

Library of Congress Cataloging-in-Publication Data is available for this publication.

ISBN: 978-0-8330-8495-8

Cover: U.S. Navy photo by Petty Officer 1st Class David McKee

RAND OFFICES

SANTA MONICA, CA • WASHINGTON, DC

PITTSBURGH, PA • NEW ORLEANS, LA • JACKSON, MS • BOSTON, MA

CAMBRIDGE, UK • BRUSSELS, BE

www.rand.org

Preface

The RAND National Defense Research Institute (NDRI) examined Department of Defense (DoD) depot-level reparable (DLR) supply chain management to assess how it could be improved to enhance customer support and reduce costs. Our research team employed complementary approaches, including analysis of DLR flow and inventory data, interviews and site visits, reviews of service documentation, a literature review, and case studies of specific DLRs. From these multiple methods, we distilled the most common reasons for apparent inventory "excess" and customer support shortfalls, and we identified associated process improvement opportunities.

We did not find any large, "silver bullet" solutions, concluding that DLRs are managed relatively well by the services. However, we did find a number of modest opportunities for improving DLR supply chain management. The first, and likely largest, is improving parts supportability, including taking a total cost perspective that encompasses supply and maintenance when planning inventory in support of depot production. A second opportunity is to shift the Army more toward pull production. A third is to reduce lead times for all types of contracts affecting DLR supply chain management. And a fourth is to better account for all resource lead times in planning DLR production and anticipatable shifts in procurement and repair needs. These enhancements would all improve customer support, with better parts support likely reducing maintenance costs and pull production reducing the buildup of inventory. Additional cost-saving opportunities are more limited because what on the surface appears to be substantial inventory excess and high disposals of assets is actually a reflection of the fact that DLRs are durable assets very much like weapon systems and other end items. We found that most DLRs have very low condemnation rates. So when they are replaced by upgraded versions or weapon systems are phased out, demands disappear but the assets remain, leading first to "excess" inventory and then disposals. This is a cost of doing business. Most DLRs get repaired many times before they become obsolete, with each asset purchased for inventory being used many times.

This project was sponsored by the Assistant Secretary of Defense for Logistics and Materiel Readiness. It should be of interest to logisticians across the DoD, financial managers, Congress, and other stakeholders interested in understanding DoD inventory management and ensuring it is executed as effectively as possible. The research was conducted by the Acquisition and Technology Policy Center of the RAND National Defense Research Institute (NDRI). NDRI is a federally funded research and development center sponsored by the Office of the Secretary of Defense, the Joint Staff, the Unified Combatant Commands, the Navy, the Marine Corps, the defense agencies, and the defense Intelligence Community. For more information on the RAND Acquisition and Technology Policy Center, see http://www.rand.org/nsrd/ndri/centers/atp.html or contact the director (contact information is provided on the web page).

Contents

Figures

Tables

Summary

The Department of Defense (DoD) has a broad array of weapon systems and other major end items with many expensive components. For many of these components, when they have to be replaced, it costs less to replace them with a repaired, refurbished spare part than to buy a new one. Such items are called *reparables* within DoD. DoD designates different levels of maintenance to conduct such repairs depending on the skill level, tooling, and facilities needed to execute the repairs, with depot-level repair representing the most sophisticated level. Reparables for which all or some repairs require this level of capability are called depot-level reparables (DLRs). Within DoD, the services manage almost all reparables, with the Defense Logistics Agency (DLA) managing consumable spare parts.

DLR inventory comprises the bulk of DoD secondary item inventory in terms of dollar value. There was an average of $100 billion in service-owned secondary item inventory on hand in fiscal year (FY) 2011, of which we estimate about $90 billion consisted of DLRs.[1] The value of inventory is not a recurring cost, though; rather, the costs associated with inventory are called *inventory holding costs*, and the assets

---

[1] Department of the Air Force, *United States Air Force Working Capital Fund (Appropriation: 4930)*, Fiscal Year (FY) 2013 Budget Estimates, February 2012; Department of the Army, *Army Working Capital Fund Fiscal Year (FY) 2013 President's Budget*, February 2012; Department of Defense, *Defense Working Capital Fund, Defense-Wide Fiscal Year (FY) 2013 Budget Estimates Operating and Capital Budgets*, February 2012. Department of the Navy, *Fiscal Year (FY) 2013 Budget Estimates: Justification of Estimates Navy Working Capital Fund*, February 2012. The $90 billion estimate is based on the percentage of service-owned inventory held in DLA distribution centers that we identified as DLRs.

themselves are a sunk cost. Obsolescence is the primary component of DoD inventory holding cost. From 2005 through 2012, the services disposed of an average of $5.1 billion of condition code "F"—unserviceable but economically repairable—and $1.4 billion of serviceable DLRs per year (valued at standard price).[2] This represents $6.5 billion (less the surcharges for supply chain management) of assets that were purchased and then later disposed of with useful life remaining. Two questions arise: (1) What led to the development of excess inventory culminating in the disposal of useable assets, and (2) could this level of excess inventory buildup be reduced?

In addition to accounting for a substantial amount of inventory, DLRs have high back-order rates with respect to wholesale-level requisitions. However, this does not directly relate to readiness because the services primarily use tactical or retail inventory, replenished by wholesale, to meet readiness goals, with most DLR wholesale back orders replenishing retail inventory, which does have to be higher to support readiness while accommodating wholesale back orders. Additionally, the service supply management organizations prioritize the release of wholesale assets based on readiness needs, including the mission priority of units. Nevertheless, to the extent that demand and supply could be more tightly aligned, the existing level of inventory could provide improved customer support. This raises two more questions: (1) What leads to the high back-order rate, and (2) could this rate be reduced?

Project Overview and Objectives

These observations and questions led to a RAND National Defense Research Institute (NDRI) project sponsored by the Assistant Secretary of Defense for Logistics and Materiel Readiness (ASD(L&MR)) to identify how the services' supply planning organizations can improve how they manage their DLR supply chains in order to (1) reduce the

---

2 DLA Materiel Information Systems (MIS) Issues file. Disposals are indicated by A5J (issues to disposal) transactions with condition codes of A, B, C, or D for serviceable item disposals.

generation of excess DLRs that are later disposed of, (2) reduce the level of inventory needed to effectively support customers, and (3) improve customer support.

To accomplish this project, the RAND research team pursued complementary approaches. Data analysis of DLR demands, inductions, repairs (measured by receipts of serviceable DLRs from maintenance), disposals, condemnations, and inventory at the Defense Logistics Agency (DLA) distribution center (DC) level enabled us to gain an understanding of some planning and supply chain management practices. Interviews, site visits, and service documentation enabled us to gain an understanding of processes, with questions informed by the data analyses and then interviews in turn helping us better understand the patterns and data. The data were also used as the basis of case studies of specific items, which were developed and used to identify the root causes of excess and shortages in conjunction with interview findings. Based on relative volumes of DLR transactions and inventory, we limited our examination to the Air Force, Army, and Navy.[3] Using these research methods, we distilled process improvement opportunities and characterized their likely limits.

Findings from Item-Level Case Studies

We selected a sample of items from across the three services we considered, targeting some that have excess inventory and others that exhibit customer support shortfalls. We then identified the item managers or others familiar with the histories of the items and set up interviews to uncover the stories behind their data patterns. For the items with apparent inventory excess, two consistent, closely related causes arose as the likely primary causes of this situation. For the shortage items, a wide variety of causes arose that are likely representative of typical

[3] For example, in 2011, there were over 200,000 issues each for DLRs managed by the Air Force, Army, and Navy, but only 3,000 for DLRs managed by the Marine Corps. Source: Logistics Response Time (LRT) database.

shortage situations, but the sample size was too small to indicate the likely distribution among these causes.

Large Disposals of Unserviceable Assets: A Cost of Doing Business Resulting from DLR Phase-Outs and Low Washout Rates

The case studies, supported by our broader process interviews with the service supply management organizations, indicate that the disposal of assets is essentially a cost of doing business when using durable assets that are economically efficient to repair and have low "washout" rates (i.e., they can be successfully repaired at an efficient cost level all or most of the time). In other words, these durable assets are simply being disposed of once the DLR is obsolete or used at a much lower rate. These DLRs have typically been repaired many times over the course of their useful lives, staying in the system once bought. When an end item is replaced or otherwise phased out, the associated DLRs are no longer needed. Similarly, during the course of an end item's life span, when a DLR is replaced by a new version to improve capabilities, to increase readiness by decreasing the replacement rate of the item, or to reduce costs by improving durability and thereby decreasing the replacement rate, the replaced DLR's inventory is no longer needed unless the DLR itself can be upgraded to the new configuration through the replacement of one or more internal components.

In the case studies, the phase-out items all had 0 or close to 0 percent condemnation rates (i.e., when inducted into repair, the depot can almost always repair them), so the assets cannot be gradually phased out through attrition as they are used and fail. Rather, as demand declines and then disappears, the assets become excess. Our interviews, combined with these data, strongly suggest that the primary reason for DLR excess is item phase-outs, combined with very low condemnation rates. In a broader data analysis, we found that the low condemnation rates for the case study items are common for DLRs, with over 80 percent having a 0 percent condemnation rate.

As durable assets, the obsolescence of DLRs should be thought of differently than how obsolescence is typically included in inventory holding cost. DLRs are much like the durable end items—or capital assets in a private sector business—they are used on, which are kept until they are no longer needed to conduct DoD missions or until they are replaced by a better end item. They should be thought of in the same way as phased out weapon systems. DoD officials, and others, would generally not measure inventory turns for these items; this inventory is not viewed as excess or waste, but as items past their operationally useful lives. For example, the Army has fields of such obsolete end items at Sierra Army Depot, there are thousands of mothballed aircraft from across the services parked at Davis-Monthan Air Force Base, and the Navy has decommissioned ships anchored or berthed pier-side at a number of locations.

Case Study Findings for DLRs with Supply Shortages

While there is a single driving cause category—DLR phase-outs—associated with DLR inventory excess, in the inventory shortage case studies, which are associated with degraded customer support, the causes were much more distributed. Still, one cause stands out. The most common factor was one or more back-ordered parts (typically DLA-managed) needed to complete repairs. Additionally, our interviews at the maintenance depots highlighted parts supportability issues as by far the greatest constraint on their production agility, and maintenance personnel reported high numbers of cases for which they were awaiting parts. At the Air Force Air Logistics Centers and Navy Fleet Readiness Centers for aviation depot maintenance, this issue was the first thing interviewees wanted to discuss because they considered it their most critical barrier to improving support and efficiency. Capacity, repair process flexibility (tooling and labor), and funding were generally not considered to be constraints on repair flexibility in the timeframe of the study. However, it is possible that in the new fiscal environment funding will become a constraint.

Other root causes identified include

- repair contract renewal delays creating gaps in production
- shortages of assets during phase-in/fielding, with different underlying problems
- anticipatable changes in repair factors that were not planned for
- low-demand, high-demand–variability DLRs producing challenging forecasting and planning.

These causes present potential process improvement opportunities associated with improving parts supportability, reducing contract lead times, establishing integrated repair planning that considers all resources, and anticipating and planning for knowable shifts in demands and condemnations. In addition to reflecting process issues that impede customer support, these case studies of shortages correspond to process issues that lengthen repair cycles and affect maintenance agility, increasing inventory requirements to meet customer needs. So improvements will reduce costs as well as improve customer support.

Implications for Measuring Inventory Turns to Monitor Inventory Management Efficiency

The case study findings have significant implications for how inventory turns, a good measure of inventory efficiency if used well, should be measured for DLRs. Including all DLRs would not present a good picture of process performance because the total population consists of several subpopulations with very different but explainable levels of inventory turns. In particular, the total DLR inventory includes substantial inventory of items that are currently being or have recently been phased out, producing very low turns for these durable assets. Combining the subpopulations does not produce a meaningful picture. Accordingly, we developed an approach for categorizing DLRs into different life cycle phases: (1) new item Phase-In, (2) Steady State usage, (3) Phase-Out, and (4) Other. The Other category includes items with no demand, which may have been phased out before the start of our data set, and items with very low, sporadic demand.

After categorizing all DLRs, we then computed turns by category and service for each category. These calculations revealed that inventory turns for steady-state items range from 0.6 to 1.4 among the services, as shown in the upper left-hand graph of Figure S.1. In other words, each DLR held in inventory goes through an entire closed-loop use cycle every 7–17 months, depending on the service. Phase-out items naturally have much lower turns, and items in the Other category have even lower turns. Notably, while most demand is in the Steady State category (see the upper right-hand graph of Figure S.1), the Phase-Out and Other categories include half the inventory (lower right-hand graph of Figure S.1), and the majority of items are in the Other category (lower left-hand graph of Figure S.1). We also see that phase-in items have high turns, indicative of one of two situations: (1) a greater propensity of customer support difficulties than excess inventory for items being phased in or (2) item upgrades that are being phased in gradually over time. In the latter case, inventory on the shelf can be limited in accordance with the phase-in plan.

This nuanced perspective in terms of measuring performance is important. DLR supply chain management process improvements, which should be monitored using a combination of metrics, including inventory turns (efficiency) and customer service metrics, will primarily affect the amount of inventory needed in steady state management. Thus, the inventory turns measurement should be limited to those items in the steady-state phase of their life cycle.

Improving Parts Supportability

Parts supportability issues impede the ability of DLR supply chain managers to provide effective customer support. Additionally, our visits to maintenance depots across the services revealed a number of common practices in the depots for dealing with or that stem from shortages for production that increase maintenance costs. Examples include cannibalization of work-in-process assets, induction of carcasses solely for the salvage of parts, and local fabrication in lots of one or small batches. Data systems do not capture these events, nor do

Figure S.1
Inventory Turns, Demands, Items, and Inventory by Life Cycle Phase

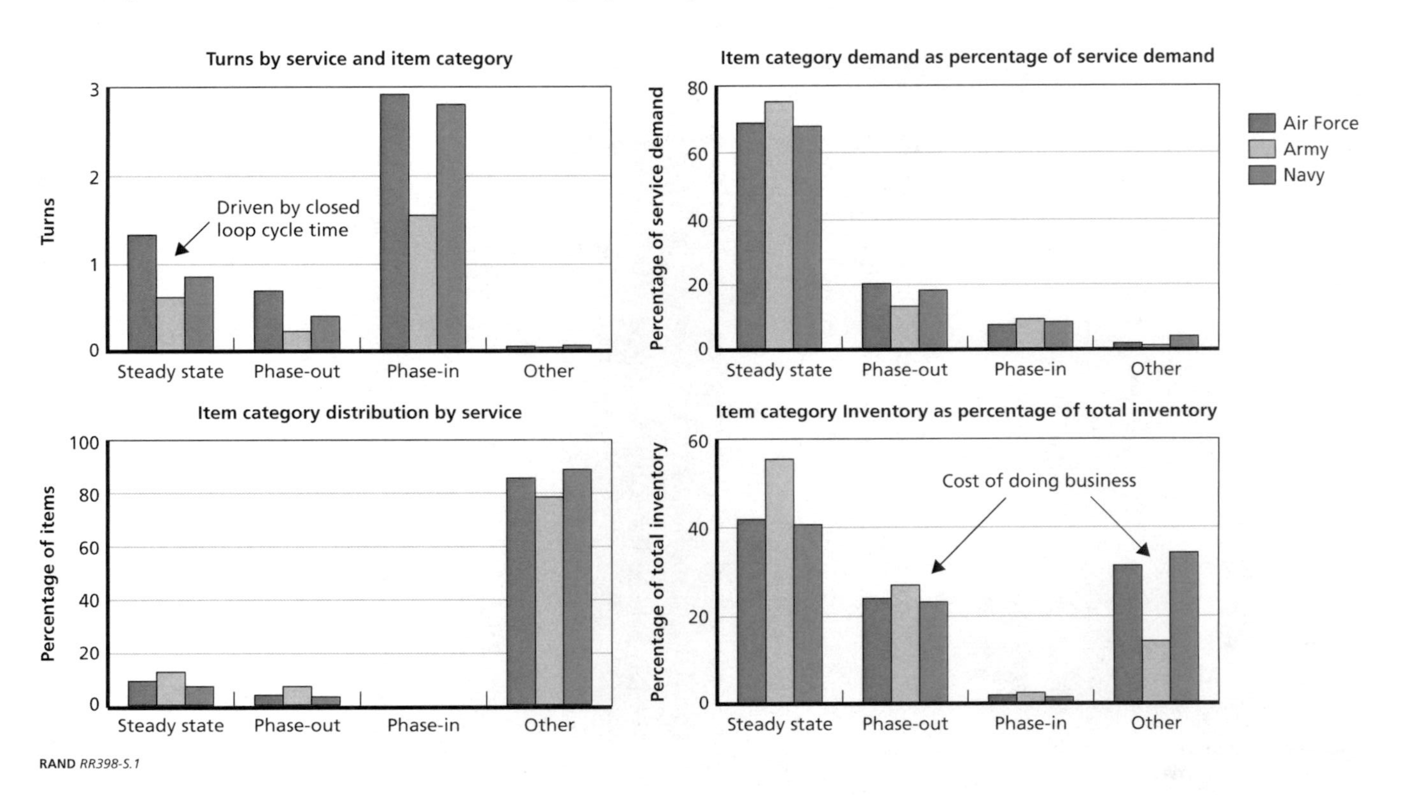

RAND *RR398-S.1*

they capture the amount of work involved in each. Thus, quantifying the costs is not possible. However, at several depots this issue is of great concern and their local data snapshots indicate substantial amounts of work affected by parts shortages, suggesting that the costs could be substantial. The services are cognizant of this issue, placing substantial emphasis on improving parts supportability. With the transition of depot maintenance parts management to DLA as a result of the 2005 Base Realignment and Closure legislation, this has introduced increased need for effective cross-organizational coordination. The services are taking an increasingly proactive role in improvement, both by focusing on what they can do to improve the situation and by being demanding customers.

One area of emphasis in the Air Force is the increased use of demand history adjustments (DHAs). When a part is not available and the depot does not submit an order for the part, executing a workaround instead, a demand record is not created. This impacts the demand history that DLA relies on to forecast future demands as part of inventory and supply planning. So the workarounds that maintenance executes can thus impede the improvement of support to these operations. DHAs allow the depot to submit a demand record that gets integrated into an item's demand history and thus into DLA demand planning. The Navy has been the biggest user of DHAs, while the Army makes limited use of DHAs.

A second area of attention focuses on improving collaboration through improvements in the use of demand data exchange (DDE). The services can use DDEs to indicate planned changes in demand from historical levels. Recognizing the need to be proactive in this regard, the Air Force developed standardized DDE generation and submission practices and codified them in a new Air Force instruction in 2012. The Navy and Army are taking a different approach, enabled by their adoption of SAP for their enterprise resource planning (ERP) systems in conjunction with DLA's adoption of the same ERP. They are electronically tying their ERPs to DLA's ERP to directly transmit parts needs to DLA that are associated with their production plans.

Accounting for the Maintenance Costs from Parts Shortages

Beyond these efforts to improve information flow to DLA to improve planning, the consequences of parts shortages suggest another path to improvement for DoD. Service-level goals for DLA safety stock are based on service-level agreements with the services that are loosely based on readiness needs and historical levels of support performance. In the private sector, when there is a stockout, there is a risk of a lost sale. In the military, the benefit of having an item in stock is readiness, which is difficult to turn into "revenue" value to trade off against inventory when considering stockout costs versus inventory costs. However, there is a monetary value component to stockouts in support of depot maintenance—additional maintenance costs. While these costs could be substantial, their amounts are unknown. Much more in-depth examination than possible in this project would be necessary to determine their levels. With the costs completely unknown, they are not considered in planning inventory levels. In contrast, if there was a reasonable estimate of the stockout cost, this could be considered in tandem with safety stock investment cost to determine the lowest total cost level of safety stock.

Parts Supportability Metrics and Performance-Based Agreements

A review of the services' performance-based agreements (PBAs) with DLA reveals limited focus on the maintenance or DLR-level perspective in the PBAs, with significant differences in PBAs across the services. The Air Force, Marine Corps, and Navy do share a relatively common core of item-level supply chain performance metrics in their PBAs that are focused on response time and item availability. However, the Air Force adds several metrics that provide maintenance and DLR perspectives and that are also intended to hold the Air Force accountable for how it affects collaboration and consumable item lead times. In addition to using this broader suite of metrics, the Air Force has worked with DLA to ensure the metrics get high-level attention. As a result, DLA includes metrics that are important to the Air Force in its monthly Agency Performance Review (APR) meetings that are held by the DLA director and attended by the director's direct reports and the rest of the DLA senior leadership team.

The Use of Pull Production

Aligning supply with demand is a hallmark of pull production. We aimed to determine the degree to which each service employs a pull production system and whether there might be value in shifts in degree for any of the services.

The Air Force system can be described as very close to a complete pull system, with daily repair induction planning based on demands, priorities, and retail inventory levels being the basis of its system. Within the Air Force, there has been some debate about whether daily planning introduces excessive turbulence into maintenance processes by degrading productivity and increasing total cost. These discussions occur over the backdrop of an underlying view that daily planning is either the standard practice in the Air Force or should be. However, during our site visits and interviews we learned that the different Air Logistics Complexes (ALCs) and shops within them, including some that technically use daily induction planning, are employing different practices to apply limited level-loading for maintenance efficiency, while still reflecting a pull production paradigm.

The Navy employs a combination of pull production and what we term *modified pull production*. Over half of the Navy's DLR repair is done at intermediate-level maintenance. These activities operate on a pull system with one-piece flow. If the DLR cannot be repaired at the intermediate level, then it is sent to storage to await induction into depot maintenance. Depot maintenance receives six-month induction plans at the start of each FY, and then two three-month or one six-month plan for the second half of the FY. For each of these periods, the depot maintenance activity determines the detailed schedule within the period, balancing production efficiency and customer needs if shortages of an item have been communicated to them. This semiannual/quarterly workloading appears to produce a serviceable/unserviceable profile similar to that of the Air Force. The result is that this frequency of planning appears sufficient to represent a pull system.

The Army process appears closer to a push system in design, with annual workloading provided to the depots and changes made by exception during the year. This tends to limit changes to the most severe cases and is a less reliable process for making all the changes that are needed. A modified pull system requires a new workload plan on a periodic basis—production will stop otherwise—and a complete pull system will only produce on demand. In contrast, a push system requires that an exception decision be made on every item for which demand has changed significantly—production will continue as planned unless altered (note that Army production will also stop if a new FY workload plan is not provided and funded). Modified pull forces a decision on every item at the frequency of the planning horizon. This is the fundamental difference between push and pull: whether changes to the plan are made by exception or are built into the process on a systematic, more frequent basis. As a result, we observe more Army items with overproduction, along with some underproduction. This is relatively limited, though, in times of stable demand. The effect becomes greater in times of shifting or highly variable demand.

A push system is most likely to lead to problems in a period of shifting demand. For example, when demands increased in FY 2003 with the start of Operation Iraqi Freedom (OIF), material availability plummeted, taking until FY 2006 to recover to close to the target at the time of 85 percent. In FY 2003 and FY 2004, production fell far behind demand for some National Item Identification Numbers (NIINs).[4] Besides leading to back orders, this led to situations in which repair capacity was insufficient for serviceable inventory to recover to a position that would provide effective customer support. This led to decisions to buy more serviceable assets to improve support. This

[4] This came partly from planning, with lags in forecast increases, and partly from a substantial delay in the flow of broken DLRs from Iraq back to depots in the United States for repair.

became a wartime necessity but increased long-term inventory, potentially increasing disposals in the future.

Balancing Productivity and the Degree of Pull Production

While aligning supply and production with demand reduces inventory overall and serviceable inventory for DLRs and can improve customer support when demand shifts, it can also have implications for maintenance costs. Depot maintenance can be executed more efficiently when production is level for each production line, enabling high capacity utilization. High capacity utilization can be combined with pull as well, but this will generally require prioritization in terms of what items to induct and thus some level of back orders in the face of demand variability. And in some cases, batch production can be more efficient. Additionally, batch production can facilitate parts supportability by enabling the ordering of shortage items to support production farther in advance. This may explain the lower levels of concern about parts supportability reported in Army depots than in Air Force and Navy facilities.

Thus, shifting to a complete on-demand pull system or daily planning is not necessarily the best total system solution. Some amount of level-loading can be important for efficiency. The degree of level-loading or push versus pull that provides the best overall result depends on a number of factors, including repair flow time, the range of items made on a line or at a work station, the flexibility of tooling and labor, the space available for work-in-process inventory, capacity utilization, and parts support lead time.

Moving Toward Pull Production for the Army

Prior analysis for the Army suggests that much of the gains from shifting to pull production could be garnered through three-month planning horizons, or even six-month horizons. This happens to be similar to current Navy practices.[5] These periods are long enough to accommodate planning windows for most parts and allow for some production smoothing while still reacting in one-fourth to one-half the time as annual workloading, and changes in workload plans would not need to rely on exception management. Thus, we suggest the Army adopt a modified pull production system with the aim of increasing workloading frequency and reducing the time horizons of firm orders passed to the depots. It could start by shifting all production planning to six-month intervals. After a period of learning and working through resulting issues, it could further decrease the intervals on a customized basis tailored to item-level demand, item-level repair profiles, and production facility characteristics.

Contract Lead Time Reductions

Efforts to reduce contract lead times would improve DLR supply chain management in three different ways. First, the longer the procurement lead time for new DLRs, the greater the potential inventory excess and customer support impacts of unanticipated changes in demand. The longer the lead time, the greater the risk of a demand shift from the plan or forecast within the lead-time window. Second, a confluence of factors can lead to long repair contract lead times to lead to excessive assets or poor customer support. Third, lead times for piece parts used in repairs can affect awaiting parts time in production and production efficiency.

Improved Repair Planning

Changes in the demand for repair, either due to demand shifts or condemnation rate shifts, can occur for anticipatable reasons that are not automatically captured in DLR management information systems and

[5] Unpublished research by Mark Wang, Jason Eng, Rachel Rue, and Jeffrey Tew on Adapting Army Secondary Item Planning to Pull Production, November 2009.

that are not always accounted for by item managers. There is a small subset of DLRs that can only be repaired a fixed number of times before the assets can no longer be renewed, leading to automatic condemnation and disposal. If this occurs in a somewhat narrow window across the associated end item fleet, condemnations can abruptly spike upward and prevent effective support through repair only. If the item manager does not anticipate and plan for this, customers can be left with poor support over the lead time for buying new assets.

A similar situation can occur early in a DLR's life cycle. If a DLR is phased in quickly and it is a wear-out type item, demands will be low until the first wear-out cycle is reached. Demands will then suddenly increase, potentially leading to resource shortage problems in support of production. This situation or any other that temporarily eliminates the need for repair can also produce a procurement delay problem for items with a relatively high condemnation rate. If no repairs occur over a substantial period, the planning factor for condemnations can be off once repairs start.

More generally, when repair has been temporarily halted or is below the ongoing level of demand, the system can be left short of assets to reinitiate repair altogether or at the higher level of demand. The maintenance planning systems use repair lead time, comprising shop flow time, to determine when to start inducting assets to have them in serviceable condition when needed. The systems also use procurement lead time to determine when new buys might need to be placed in the face of demand increases or condemnations. Repair requires labor and parts. If these are not on hand when repair induction is called for, repair cannot start or be completed. The planning system ideally would anticipate the need for these resources a lead time before the needed induction date. This could be done in two ways. One way would be to have the lead time for all supporting resources recorded in the planning system and have the resource with the longest lead time produce the needed action initiation lead time. The second way is for the item manager to manually determine this and plan accordingly, such as developing DDEs for DLA to purchase needed parts in anticipation of production.

Overall Conclusions, Recommendations, and Needs for Further Research

DLR supply chain management appears to be done relatively effectively across the services. In particular, there does not appear to be any single process improvement opportunity for dramatically reducing inventory requirements. However, improving the processes for providing needed parts to depot maintenance would improve customer support and could reduce total system costs through improved maintenance productivity.

It is difficult to find items with excess inventory from avoidable situations or poor supply chain management decisions or processes. Instead, the buildup of excess and the resulting disposals stem primarily from DLR phase-outs due to upgraded replacement DLRs and end item fleet size reductions and phase-outs combined with very low condemnation rates. DoD should better explain the item phase-out impact on DLR inventory for improved understanding by external stakeholders and should isolate excess DLR inventory and disposals to make them visible and distinct within overall inventory and disposal reports.

When items are replaced or end-item fleets are phased out, the plans should include the long-term disposal plans for the associated DLRs. The data indicate that in most cases of excess leading to disposals, there is a multiyear delay before disposals occur. It is likely that these delays could be shortened, reducing storage costs and, potentially, warehouse infrastructure requirements. This could also improve perceptions of DoD inventory management by increasing turns, reducing excess, and reducing overall inventory. Similarly, when looking at inventory turns, phase-out items should be separated out to provide a better understanding of turns for steady-state items. Focusing on the steady-state items will provide a better understanding of how closed-loop DLR process improvement would affect inventory requirements for new DLRs.

Beyond this, the one major practice gap in DLR supply chain management that hinders aligning supply with demand to provide effective customer support and avoid building up excess is the Army's use of a push-like production system. The Army should take steps to move toward a more pull-like paradigm. More broadly, the services should seek to find the workload planning horizons that minimize total costs when considering supply and maintenance, seeking to shorten the horizon without impacting maintenance productivity. Shorter horizons enable faster response to changes in demand and reveal process inefficiencies.

A broader issue that impacts DLR supply chain management effectiveness is parts supportability. The biggest direct impact of this issue is degradation in customer support when severe parts shortages lead to the complete draining of serviceable wholesale inventory of a DLR. Additionally, parts delays and variability in wait times increase the closed-loop repair cycle, thus increasing inventory requirements. They can also influence service selection of planning horizons (i.e., where the service selects to be on the push-versus-pull spectrum) and thus the responsiveness of the DLR supply chain to shifts in demand, further influencing inventory and customer support. Finally, parts supportability issues impact maintenance productivity and costs through the various workarounds executed by the depots to compensate for parts shortages. Such costs could be substantial but are not directly measureable given the data that are currently collected.

There are several complementary paths to improved parts supportability and supply chain integration. One would be for the services to quantitatively estimate the costs of parts shortages and for DLA to incorporate these costs in safety stock planning to jointly trade off shortage and inventory holding costs from safety stock to determine the balance that produces the lowest total cost for DoD as a whole. The second is to continue improvements in collaboration and coordination. Effective service processes for sharing planning data with DLA in actionable form are key. This includes both automated data sharing, as feasible, and the manual provision of actionable information. In conjunction, having the right metrics and using them for PBAs between the services and DLA, with the Assistant Secretary of Defense for Logistics and Materiel Readiness (ASD(L&MR)) establishing a base template, would provide a communications tool and better ensure that interorganizational priorities are aligned, driving effective execution of both collaborative and intra-organizational planning processes.

Finally, when the planning processes that warn of repair-plan changes break down or unexpected changes in demand occur, reduced lead times for procuring piece parts can reduce the time spent awaiting parts. Long lead times also impact DLR customer support effectiveness and efficiency in three other ways. This first is when a long repair contract lead time leads to a lapse in having a repair contract in place or prevents the timely initiation of a newly needed new repair contract. Second, if a demand increase requires an increase in total inventory of a DLR, the lead time for procuring new DLRs could affect customer support, depending on the demand increase warning lead time and how it compares to the procurement lead time. Third, if there is a planned increase in demand that triggers increased buys of both a DLR and its indentured parts, but the plan does not materialize, this will lead to excess inventory. The longer the lead times, the greater risk there is of this occurring.

In addition to parts planning improvements, there is some potential for repair planning improvements. The service planning systems determine when repairs need to be initiated based on the repair flow time. This assumes that repairs can be initiated upon induction and that there will not be delays caused by supporting resources. However, particularly when an item has had a gap in production (or when it has not had a repair program before), not all of the requisite resources are always immediately available. If lead time is required to get these in advance of starting a repair, the systems need to flag this as well, kicking off the purchasing process for parts or a repair contract a lead time in advance. Ideally, this would be automated in the planning systems. But until this is possible, item managers should check for these conditions periodically for items not currently in production, looking ahead to when production is projected to need to start, checking the supporting parts inventory, and checking for the existence of a repair contract, as appropriate to the specific DLR. Similarly, item managers need to pay special attention to new items and lifetime replacement–limited items, tracking installation periods and the number of lifetime replacements, respectively. From this, they should track when they would expect demands and condemnations, respectively, to increase.

The case studies we conducted for this research were essential for identifying causes of DLR inventory excess and customer support shortfalls. However, the number of case studies that could be coordinated and accomplished as part of the project was limited to a couple dozen. A larger sample could provide further information on the root causes of process problems, and as our recommendations or other process improvements are implemented, new case studies could help provide feedback on process change effectiveness. So we recommend that the NIIN case study approach be employed as part of future process improvement efforts.

Additionally, certain situations could be automatically flagged when they occur to identify valuable cases for analysis. For example, if a critical customer support shortfall forces a procurement action when significant unserviceable inventory of a DLR is on hand, it could be recorded and could be a trigger to identify the constraint on production that forced the purchase of additional inventory. The resulting accumulation of data on production constraints could lead to the identification of key process shortfalls hampering system efficiency.

Beyond the process disruption events and overall production planning approaches that are the focus of this report, the services have long sought to improve DLR processes, focusing to a great degree on repair process flow time, and, in some services, retrograde time. This reduces the amount of inventory the closed-loop system needs to meet customer needs, which reduces initial buy requirements. In turn, when an item is phased out, this will be reflected in reduced disposals of economically useful assets. Such process improvements should be ongoing in the spirit of continuous improvement. In addition to likely improving maintenance efficiency, this will reduce DLR purchase requirements each time a new DLR is phased in or a new system is fielded.

Acknowledgments

The Honorable Alan Estevez, as Assistant Secretary of Defense for Logistics and Materiel Readiness, seeking to better understand the factors affecting inventory and customer support levels and find opportunities to improve these levels, sponsored this research. Mr. Paul Peters, first as Deputy Assistant Secretary of Defense (DASD) for Supply Chain Integration and then Principal DASD for Logistics and Materiel Readiness, helped to shape the direction of this research and provided guidance throughout the project to ensure its success.

Throughout the course of this project, we received exceptional support from the leaders, staff, technical experts, and line personnel in the supply chain management and maintenance organizations within the Air Force, Army, and Navy. We thank them for their willingness to devote their time to explaining processes and reviewing case study histories to help identify process improvement opportunities and enable us to better explain the cost structure of DLR supply chain management.

Independent reviewers, Lieutenant General (retired) Mitchell Stevenson and Louis Miller of RAND, provided constructive comments that helped us sharpen the message of this report and refine the supporting information. Within RAND, John Schank and Mark Arena coordinated visits with key Navy offices and conducted several interviews and process visits. Dan Romano, through his role as a RAND liaison at Air Force Materiel Command, facilitated connections for interviews and information requests and assisted with understanding the command's structure. Ken Girardini, as the Military Logistics Program Director, provided insightful comments that enabled us to refine the report. Laura McMillen helped prepare the document.

Abbreviations

AFMC	Air Force Materiel Command
AFSC	Air Force Sustainment Center
AFTC	Air Force Test Center
ALC	Air Logistics Complex
AMC	Army Materiel Command
AMCOM	Aviation and Missile Life Cycle Management Command
ANAD	Anniston Army Depot
AOIB	Army Organic Industrial Base
APR	Agency Performance Review
ARNG	Army National Guard
ASA FM&C	Assistant Secretary of the Army for Financial Management and Comptroller
ASCC	Army Service Component Command
ASD	Aviation Support Division
ASD(L&MR)	Assistant Secretary of Defense for Logistics and Materiel Readiness
AVCAL	Aviation Consolidated Allowance List

AWCF	Army Working Capital Fund
AWP	awaiting parts
AWPS	Army Workload and Performance System
BOM	bill of materials
BRAC	Base Realignment and Closure
CCAD	Corpus Christi Army Depot
CECOM	Communications and Electronics Life Cycle Management Command
CNRMC	Commander Navy Regional Maintenance Center
COG	cognizance
COMFRC	Commander Fleet Readiness Centers
COSAL	Consolidated Ship Shipboard Allowance List
DASD	Deputy Assistant Secretary of Defense
DASD(SCI)	Deputy Assistant Secretary of Defense for Supply Chain Integration
DC	distribution center
DDE	demand data exchange
DHA	demand history adjustment
DLR	depot-level reparable
DLA	Defense Logistics Agency
DMOPS	Depot Maintenance Operations Planning System
DoD	Department of Defense
DOL	Directorates of Logistics

DP	demand planning
ERP	enterprise resource planning
EXPRESS	Execution and Prioritization of Repair Support System
FDD	forward distribution depot
FRC	Fleet Readiness Center
FY	fiscal year
GCSS-A	Global Combat Support System–Army
HQ	Headquarters
ICP	inventory control point
ICRL	Individual Component Repair List
IM	item manager
IMSP	Inventory Management and Stock Positioning
KPSS	Kwiatkowski-Phillips-Schmidt-Shin
LCMC	Life Cycle Management Command
LEAD	Letterkenny Army Depot
LMP	Logistics Modernization Program
LRT	Logistics Response Time
MCC	Materiel Control Code
MICAP	Mission Impaired Capability Awaiting Parts
MIS	Materiel Information Systems
MRP	material requirement planning
NAVAIR	Naval Air Systems Command

NAVICP	Navy Inventory Control Point
NAVSEA	Naval Sea Systems Command
NAVSUP	Naval Supply Systems Command
NDRI	National Defense Research Institute
NIIN	National Item Identification Number
NLCO	National Logistics Coordination Office
NMP	National Maintenance Program
NSFLMD	National Sustainment and Field Level Maintenance Division
NSN	national stock number
NWCF	Navy Working Capital Fund
OC-ALC	Oklahoma City Air Logistics Complex
OEM	original equipment manufacturer
OIF	Operation Iraqi Freedom
OMA	Operations and Maintenance Army
OMB	Office of Management and Budget
OPS	Operational Program Summary
OPTEMPO	operating tempo
ORT	order response time
PBA	performance-based agreement
PBOM	production bill of material
PCF	program change factor
PDM	programmed depot maintenance
PDMC	Planning for DLA-Managed Consumables

PM	program manager
POM	Program Objective Memorandum
QBO	quantity by owner
RAPS	Rotables Allocation and Planning System
RBOM	repair bill of materials
RC	Recoverability Code
RI	Rock Island
RMC	regional maintenance center
SARSS	Standard Army Retail Supply System
SCMG	Supply Chain Management Group
SCMW	Supply Chain Management Wing
SOR	source of repair
TACOM	Tank-Automotive and Armaments Life Cycle Management Command
TRF	technical replacement factor
TYAD	Tobyhanna Army Depot
WCF	Working Capital Fund
WR-ALC	Warner Robbins Air Logistics Complex
WIP	work in process
WSS	Weapon Systems Support

CHAPTER ONE

Introduction

The Role of Depot-Level Reparables in Sustainment

The Department of Defense (DoD) has hundreds of thousands of end items encompassing thousands of different types and models ranging from small arms to radars to tanks and aircraft. Many of these weapon systems and other end items are complex systems with many expensive components. For many of these components, when they fail or are replaced in scheduled maintenance, it costs less to replace them with a repaired, refurbished spare part rather than buy a new one. Such items are called *reparables* within DoD. Parts that are not economical to repair or cannot be repaired are termed *consumables*.

Depending on the technical skill and tooling and equipment needed for component repairs, DoD designates different levels of maintenance to conduct such repairs, with depot-level repair representing the most sophisticated level, requiring highly skilled labor and specialized tools and facilities. Reparables for which all or some repairs require this level of capability are called depot-level reparables (DLRs). These repairs require the use of consumable spare parts or even nested reparables. Consumables are also sometimes directly replaced on end items (e.g., engine oil filter).

Within DoD, the services manage almost all reparables, particularly the DLRs, for the end items they manage, with the Defense Logistics Agency (DLA) managing the bulk of consumable spare parts. Within the services, DLRs and other secondary items are managed by the Air Force Materiel Command (AFMC), the Army Materiel Command (AMC), the Naval Supply Systems Command (NAVSUP), and the Marine Corps Logistics Command (MARCORLOGCOM). Item management includes demand and supply planning, including inventory management. Supported by service-specific information systems, item managers forecast future needs based on historical demand and intelligence on planned activities and determine the inventory levels, production plans, and procurement plans needed to most efficiently meet these needs with sufficient confidence to maintain desired readiness levels. Each service has a distinct organization and process for item management, which we detail in Appendixes A, B, and C for the Air Force, Army, and Navy, respectively.

When a DLR is needed, the maintenance organization replacing the DLR on the end item typically requests the replacement from the supply system and either receives it immediately or has to wait if there is a stockout (i.e., no serviceable inventory available), returning the unserviceable DLR, or "carcass," to the supply system. Item managers direct the supply system to send the unserviceable DLR to a maintenance activity for repair when needed in accordance with their overall supply plan. Accounting for forecasted and planned demand, as well as production capabilities, their supply plans are designed to provide sufficient availability of DLRs for replacement on end items to meet readiness needs in the field and to support depot end-item programs while minimizing inventory. In short, item managers aim to meet targeted service-level goals and readiness needs as efficiently as possible.

DLR Investments, Flows, and Customer Support

Given typical DLR prices and the large breadth and quantity of DoD end items to be supported, the annual dollar value of DLR demands and the level of DLR inventory are substantial, comprising the bulk of DoD secondary item inventory in terms of dollar value. Based on the services' Working Capital Fund (WCF) budgets, there was an average of $100 billion in service-owned secondary item inventory on hand in fiscal year (FY) 2011,[1] with an estimated $64 billion (at standard price) held in DLA distribution centers (DCs).[2] We estimate that $58 billion of the service assets in DLA DCs are DLRs based on service item identification and management codes and whether the items have been repaired in depots. DLA manages the warehousing and physical distribution of these assets at the "wholesale" level, with the services maintaining ownership, stock positioning control, and distribution management of the assets. Assuming the percentage of service assets in DLA DCs that are DLRs holds for the entire service-owned, secondary-item inventory, this would produce an estimate of $90 billion of DLR inventory.

The value of inventory is not a recurring cost, though; rather, the costs associated with holding inventory are called *inventory holding costs* and the assets themselves represent a sunk cost. Typically, there are several components considered to be part of inventory holding costs. These include the cost of capital, shrinkage, storage costs, and obsolescence. Based on DLA data, storage costs are quite low, at approximately $140 million for FY 2011 across the services.[3] As of the writing of this report, the cost of capital based on Office of Management and

[1] Department of the Air Force, *United States Air Force Working Capital Fund (Appropriation: 4930)*, Fiscal Year (FY) 2013 Budget Estimates, February 2012; Department of the Army, *Army Working Capital Fund Fiscal Year (FY) 2013 President's Budget*, February 2012; Department of Defense, *Defense Working Capital Fund, Defense-Wide Fiscal Year (FY) 2013 Budget Estimates Operating and Capital Budgets*, February 2012; Department of the Navy, *Fiscal Year (FY) 2013 Budget Estimates: Justification of Estimates Navy Working Capital Fund*, February 2012.

[2] *Standard price* is the cost that a DoD customer of a WCF would pay for a serviceable item or DLR without turning in a broken one in exchange. It is comprised of the latest acquisition cost of the item plus the appropriate cost recovery rate or surcharge for the WCF organization that manages the item. Service-owned inventory on hand in DLA distribution centers was determined using the DLA Quantity by Owner (QBO) file, which records the amount of inventory by location and owner. The specific logic used to identify DLRs is described in Appendix D.

[3] FY 2011 storage costs (Department of Defense, 2012) multiplied by 50 percent—the approximate percentage of the aggregate volume of material in DLA depots that are service-managed items (based on analysis of DLA QBO file data).

Budget (OMB) guidance is also very low[4] and some argue whether this is a real cost for DoD inventory when the rate is positive. Were real interest rates to rise, OMB guidance would then call for the use of a correspondingly higher cost of capital.[5] Shrinkage is generally quite low as well. This leaves obsolescence, which is the primary component of DoD inventory holding costs for DoD inventory in general.

From calendar year 2005 through 2012, the services disposed of an average of $5.1 billion of condition code "F"—unserviceable but economically repairable—and $1.4 billion of serviceable DLRs per year (valued at standard price) from DLA DCs.[6] Figure 1.1 shows this by service for the Air Force, Army, and Navy. This represents $6.5 billion (less the surcharges for supply chain management) of assets that were purchased and later disposed of with useful life remaining, with the serviceable items representing either new, never-used items or unnecessarily repaired items. To the extent that any of these assets were not needed to effectively support customers, they would represent unnecessary expenditures and thus contribute to inventory holding costs. Several questions arise: What led to the development of excess inventory culminating in the disposal of useable assets and could this be reduced? Is this truly a component of DoD inventory holding costs, or is there some business reason driving this? In other words, how much, if any, of the $6.5 billion in annual disposals represents inventory holding cost rather than cost associated with delivering value and is thus a target of opportunity for improved efficiency?

---

[4] For an explanation, see Eric Peltz and Marc Robbins with Geoffrey McGovern, *Integrating the Department of Defense Supply Chain*, Santa Monica, Calif.: RAND Corporation, TR-1274-OSD, 2012.

[5] For guidance on how to determine appropriate costs of capital to use for government cost analyses, see Office of Management and Budget, *Guidelines and Discount Rates for Benefit-Cost Analysis of Federal Programs*, Circular No. A-94, October 29, 1992 (Appendix C, revised December 2011).

[6] Analysis of DLA Materiel Information Systems (MIS) Issues file data. Disposals are indicated by A5J (issues to disposal) transactions with condition codes of A, B, C, or D for serviceable item disposals. These figures exclude any additional disposals directly from service-managed warehouses.

Figure 1.1
Disposals of Reparables, by Service

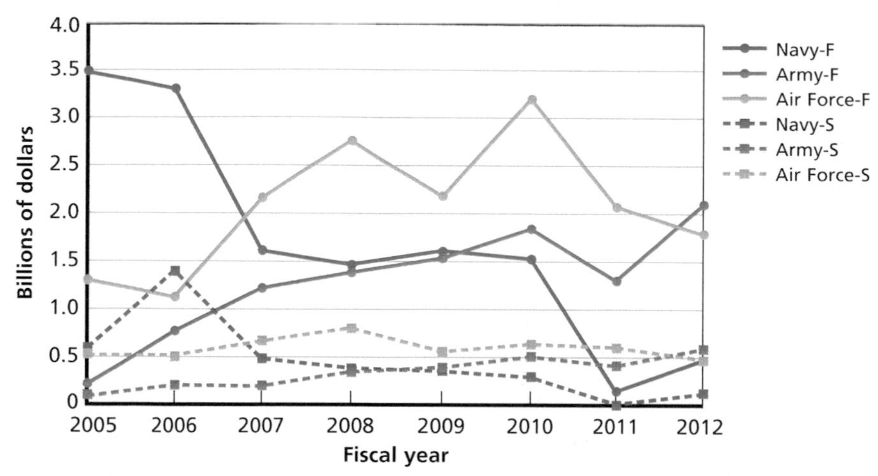

SOURCE: Analysis of the DLA MIS Issues file data and DLA QBO weekly file data.
NOTES: S = serviceable (condition codes A, B, C, and D), F = unserviceable and economically reparable. This is limited to disposals from DLA DCs, thus excluding any additional disposals directly from service-managed warehouses.
RAND *RR398-1.1*

Related to this level of disposals, in FY 2011 there was an average of about 2.1 years of supply of DLRs (serviceable and unserviceable combined) on hand across all DLRs held in DLA DCs relative to issues from these DCs.[7] The question is whether this level of inventory and the resulting inventory turns reflect the levels needed for effective customer support or whether customers could be supported with less inventory on hand, on average, and thus higher inventory turns. If so, then fewer assets would need to be purchased up front, reducing annual expenditures on DLRs for new items being phased in and later reducing disposals of useable assets.

Three factors drive the level of wholesale inventory needed:

1. the closed-loop cycle time from customer return of a broken DLR to repair completion and receipt at a DC. At this point,

[7] Sources: DLA QBO file data for on-hand and DoD Logistics Response Time (LRT) and DLA MIS data for issues.

it is then ready for issue to a customer for use on an item or to replenish relatively thin retail inventory (which is the primary source of inventory to quickly replace critical parts on end items that are not mission capable to maintain readiness).

2. the demand rate and variability
3. the service-level target (service level is typically called *materiel availability* in DoD).

A fourth factor determines whether the inventory is kept at the level needed based on these three determinants of inventory requirements and whether the service-level target is hit: the degree to which production is aligned with demand and the time lags between demands and inductions.[8] The longer the lag and the less the alignment, the more likely it is that there will be a demand change between the point in time at which production is planned and when it is executed, leading to either overproduction or underproduction. This can be "solved" by procuring more assets to meet customer demands or allowing shortages to develop. The safety level is only designed to handle typical variability around stationary demand, not shifting or nonstationary demand. Reducing this lag or the process times would reduce unnecessary buildup of inventory from such shifts and would reduce inventory shortages.

In addition to accounting for a substantial amount of inventory, DLRs have high back-order rates with respect to wholesale-level requisitions, although the effect on customer support in the field is significantly muted by retail inventory. In both 2011 and 2012, 48 percent of high-priority[9] requisitions sent to wholesale could not be filled immediately, or, in other words, were back-ordered. Combined with 57- and 47-day average wait times, respectively, for back-ordered items to be ready for issue, this produced overall average back-order times of 27

---

[8] Using DoD DLR inventory planning terms, this represents how quickly an induction occurs when on-hand serviceable inventory of an item reaches the repair action point, which is the level determined necessary to keep sufficient serviceable stock on hand to meet the service-level target.

[9] *High priority* is defined as issue priority group 1, which is intended for situations in which an end item needs a part to be returned to mission-capable status. Source: LRT data.

and 23 days, respectively, in 2011 and 2012.[10] These high-priority requisitions represented 18 percent of the total wholesale requisitions for DLRs during this period, equating to 9 percent of wholesale requisitions for DLRs being high-priority back orders. A little less than half of these, or 4 percent of wholesale requisitions for DLRs, had entries in the required delivery date field explicitly representing the need to receive the DLR to bring an end item to mission-capable status.[11] However, we cannot specifically describe the readiness impact this has because we do not have data indicating the percentage of high-priority requests, particularly those with a clear mission-capability impact, filled by retail inventory. Therefore, we cannot determine the actual back-order rate or time from a field maintenance and readiness perspective. Additionally, the service supply management organizations prioritize the release of wholesale assets based on readiness needs that encompass unit priorities, including such factors as whether or not units are deployed for operations. For example, a DLR in short supply might be rationed, with assets only released to the highest-priority units based on their current activities. Still, there is some level of customer support impact from DLR wholesale back orders, and retail inventory has to be higher to support readiness while accommodating wholesale back orders.

Project Overview and Objectives

These observations and questions led to a RAND National Defense Research Institute (NDRI) project sponsored by the Assistant Secretary of Defense for Logistics and Materiel Readiness (ASD(L&MR))

---

[10] For back-ordered requisitions, the median back-order times in 2011 and 2012 were 15 and 16 days, respectively, and the 75th percentile times were 56 and 55 days. For back-ordered requisitions placed in 2011, the 95th percentile time was 308 days. The 95th percentile time cannot yet be determined for 2012 back-ordered requisitions because of some 2013 DoD processing delays in making the data used to compute these metrics available. Source: LRT data.

[11] These were required delivery date entries of "999," which indicates Mission Impaired Capability Awaiting Parts (MICAP), or starting in N for non–mission-capable supply. Source: LRT data.

to identify how the services' supply planning organizations can improve how they manage their DLR supply chains in order to (1) reduce the generation of excess DLRs that are later disposed of, (2) reduce the level of inventory needed to effectively support customers, and (3) improve customer support. The primary value of reducing the generation of excess would be avoiding the buys in the first place; the level of reduction in annual disposals that could be achieved represents an equivalent reduction in annual spending on new DLRs. It is this reduction, if feasible, that would represent the major savings opportunity that would come from improved DLR supply chain management. Additionally, reducing the amount of inventory needed to effectively support customers, which would stem from reducing the closed-loop time from the return of a broken DLR by a customer to the return of the DLR to serviceable inventory or directly to a customer, would also reduce new buys when items are phased in and reduce the total amount of inventory to be potentially disposed of when items are phased out if their condemnation rates are low.

Improving the supply chain can be done in two ways: reacting better to changes and performing better in steady-state conditions. More tightly aligning supply and demand would enable the supply chain to more effectively respond to changes in demand, thereby reducing inventory needs and excess and improving customer support. Aligning supply with demand has two major components. The first is the planning process for the supply chain management organization encompassing such factors as planning horizons, change processes, and the degree to which push production is used based on relatively long-range forecasts versus pull production based on actual demands as they occur. The second has to do with the ability of the rest of the system to react to changes in the supply plan. This ability depends on the agility of depot maintenance (labor, tooling, and equipment), its capacity, parts support to production, funding for production and contract repair, and the consistency and speed of the return of broken items. With respect to steady-state conditions, the speed of depot repair, retrograde of carcasses of broken DLRs back to DCs, and parts provisioning processes, along with the time from receipt of a broken item to induction into repair, determine the closed-loop process time, which, in conjunction with the demand level and variability, determines the amount of inventory needed to support customers at the desired service level.

Thus, this research sought to identify process and policy changes in DLR supply chain management and how it is coordinated with production and repair parts planning to enable more-flexible depot production that is more closely aligned with demand, along with actions that would shorten the closed-loop repair cycle. To support doing so, we sought to identify best, or potentially new, practices to promulgate across the services while allowing for adaptations to their specific needs. We also sought to identify any barriers to efficiency that should be addressed in the areas of depot repair, parts support, and funding. The purposes of the changes we identified are to reduce costs over the long term and improve readiness by (1) preventing the need to increase buys to "catch up" after periods during which repair levels were too low to meet demand due to planning lags, (2) reducing the buildup of excess assets in the system, and (3) reducing stockouts.

Approach

To accomplish this, we pursued complementary approaches. Data analysis of DLR demands, inductions, receipts of serviceable DLRs from maintenance (successful repairs), receipts of unserviceable DLRs from maintenance (unsuccessful repairs or salvage operations), disposals, condemnations, and inventory at the DLA DC level enabled us to gain an understanding of some planning and supply chain management practices. Interviews and site visits, along with service documentation, enabled us to gain an understanding of processes using questions informed by the data analyses. And the interviews, in turn, helped us better understand the patterns and data. Given that it was necessary for us to understand the level of production flexibility and any constraints that hinder it, along with how external factors affect DLR repair productivity, we visited and interviewed depot maintenance in addition to the supply management organizations of each service.[12] We also used the data as the basis of case studies of specific items identified by National Item Identification Numbers (NIINs) to identify the root causes of excess and shortages, and we supplemented these case studies with interview findings. Based on relative volumes of DLR transactions and inventory, we limited our examination to the Air Force, Army, and Navy, excluding Marine Corps–managed DLRs.[13] Using these methods, we distilled process improvement opportunities and characterized their likely limits. Our research was also informed by a literature review of private sector practices.

Organization of This Report

Chapter Two summarizes the NIIN case study results, the conclusions distilled from them, and the implications these conclusions have for process measurement, understanding, and improvement opportunities. The subsequent chapters go into more depth on the process improvement opportunities, incorporating the case studies to illustrate issues as appropriate and providing information garnered through interviews, site visits, and documents. Chapter Three discusses the need for and potential paths to improving parts supportability. Chapter Four discusses where each service is on the push-pull production spectrum and the value of using pull within the context of broader process constraints and capabilities. Chapter Five covers the different

---

[12] Visits and interviews were conducted at AFMC Headquarters (HQ), NAVSUP HQ and Weapon Systems Support (WSS) in Philadelphia and Mechanicsburg (aviation and ships), Naval Air Systems Command (NAVAIR) HQ, Naval Sea Systems Command (NAVSEA) HQ, AMC HQ, the Tank-Automotive and Armaments Life Cycle Management Command (TACOM), the Aviation and Missile Life Cycle Management Command (AMCOM), Ogden Air Logistics Complex (ALC) (supply and maintenance), Oklahoma City ALC (supply and maintenance), Warner Robins ALC (supply), Anniston Army Depot, Corpus Christi Army Depot, Fleet Readiness Center (FRC) Southwest North Island, Puget Sound Naval Shipyard, and FRC Mid-Atlantic Oceana.

[13] For example, in 2011, there were over 200,000 issues each for DLRs managed by the Air Force, Army, and Navy, while there were just 3,000 issues each for DLRs managed by the Marine Corps. Source: LRT database.

ways in which reductions in contract lead time would improve DLR supply chain management. Chapter Six examines niche cases in which improved planning would improve customer support. We then shift gears in Chapter Seven, briefly summarizing the limited literature on private sector rotable supply chain management, which suggests some paths that DoD could explore for improvement. Chapter Eight pulls together all the prior chapters and delineates recommendations for the Office of the Secretary of Defense (OSD) and the services.

CHAPTER TWO

Findings from Item-Level Case Studies

The overall inventory turns and back-order rates indicate that there are many DLR NIINs with excess inventory or shortages. Manual scans of inventory and transaction time-series data at the NIIN level, supplemented by automated searches, reveal this to be the case. But these data do not indicate the reason why inventory excess and shortages occur. To determine why requires talking to the personnel who have managed the NIINs to hear the story on each one. As part of this project, we identified NIINs exhibiting excess or shortages to conduct case studies to develop a better understanding of the root causes of these issues.

To do this, we searched the data for NIINs with clear patterns of apparent excess inventory or periods of shortages impacting customer support. By month, we documented ten years (2003 through 2012) of demands, serviceable and unserviceable inventory on hand, inductions into maintenance, receipts from maintenance by condition code (serviceable, still unserviceable, or condemned), average inventory control point (ICP) time to release an order for a serviceable issue to a customer, and returns from customers.[1] Apparent excess inventory was indicated by having many years of supply on hand, often dominated by unserviceable inventory. Shortages were evidenced by periods with very low or no serviceable inventory and relatively high ICP processing or back-order times in the fulfillment of customer requests. We selected a sample of NIINs across the three services with some having excess

[1] ICP time is measured as the time from when the A0 requisition is electronically received in the Defense Automated Addressing System (DAAS) and routed to the appropriate ICP to when the A5 materiel release order is issued for the asset to be released and issued.

inventory and others exhibiting customer support shortfalls. We then sought to identify the item managers of these NIINs or others familiar with the histories of the NIINs and set up interviews to uncover the stories behind the data patterns. In all, we were able to conduct 24 case study interviews to discuss 14 NIINs with periods of excess supply and 15 NIINs with periods of shortages, with five having periods of both. The interviews for these case studies required either an item manager or a supervisor who had been in the position for several years, as these inventory and support patterns take several years to play out and indicate the types of problems we were seeking to understand.

In general, we found that the patterns we observed in the data were consistent with the internal data used by the services and the memories of the item managers and their supervisors. These personnel were able to recall the events that caused the patterns exhibited in the data, enabling the development of root causes. For the excess NIINs, there were a couple of consistent common causes that are very likely the primary causes of these situations. For the shortage items, there were a wide variety of causes, which are all likely representative of typical shortage situations, but our sample size was too small to estimate the relative distribution among these causes. Tables 2.1, 2.2, and 2.3 show the NIINs used in our case studies, whether they had excess or shortage conditions, the primary reasons for the excess or shortage conditions, and a brief summary of the NIIN's history over the case study period. The next two sections then summarize the case study results.

Table 2.1
Items Used for the Case Studies and Summaries

Service	NIIN	Excess	Shortage	Item Description	Case Study Summary
Army	014585361	X		Gas Turbine Engine	Overbuy in response to initial demands at phase-in.
Army	14426926	X		Rotary Wing Spindle Head	There is a high condemnation rate and demand is expected to shift from the full assembly to the spindle head, which would eliminate the excess, so this is a temporary situation.
Army	13799894	X		Circuit Card Assembly	Item is being phased out. The supply of this item was intended to be exhausted, but some customers/units only want the new item. This item is still being repaired.
Air Force	11894176	X		C-5 Brake Assembly	Being phased out in favor of a more durable brake assembly and replaced through attrition.
Air Force	06202517	X		Cowing, TF-39 Engine on C-5	Being phased out in 2016. Current repair contract ends in 2013—repairing enough for the remaining life cycle up front with the last repair contract.
Air Force	12348535	X		Radar Data Processor,AN-APG63 radar suite for F-15C/D and AC-130U	Excess unserviceable from the phase-out of F-15A/B. Also planned for replacement.
Navy	200041947	X		Infrared Receiver, P-3 AIMS Turret	Well managed overall. Item was upgraded for capabilities, reliability, and maintainability. Used old carcass. The higher reliability of the upgraded DLR has led to excess inventory.

Table 2.1—Continued

Service	NIIN	Excess	Shortage	Item Description	Case Study Summary
Navy	14353720	X		Helicopter Transmission, H-46	Supported end item being phased out. Some transmissions harvested from retired helicopters to salvage parts for repair, further increasing inventory.
Navy	12711063	X		Fluid Regulating Valve, FA-18	Phased out and replaced through attrition.
Air Force	145395245	X	X	KC-135 Brake Assembly	Fleet size was reduced after initial buy contract, leading to excess assets. Phased in as a reliability upgrade. Once it had to start being repaired, the DLR experienced DLA parts shortages; working on getting contract in place. There are also bill of materials (BOM) issues.
Air Force	10803407	X	X	F-15 A-D Right Landing Gear Piston	A planned change to make landing gear mandatory replacement in program depot maintenance (PDM) cancelled. Additional assets to support the change were ordered, but there was about a four-year lead time. Also, lack of repairs led to the condemnation rate history being eliminated from the information system (only recorded for two years). Parts constraints kept assets unserviceable.
Air Force	10967677	X	X	C-130 Aircraft Propeller Component	Fleet size decreasing. BOM and parts supportability issues. Excess unserviceable assets with no serviceable on shelf. High war-reserve requirement.

Table 2.1—Continued

Service	NIIN	Excess	Shortage	Item Description	Case Study Summary
Navy	13177764	X	X	Aircraft Rudder, FA-18	New, unanticipated failure condition developed late in life, requiring fleet-wide replacement.
Navy	11117804	X	X	Seal Ring Assembly	Hit life cycle limits across fleet (three repairs), which was not accounted for in planning. May have over-procured in response, creating a temporary excess condition that will be drawn down as demands continue.
Army	12844013		X	Aircraft Gas Cold Section Module	Delay in contract award and then approval of product verification article to initiate repair. Additional assets for customer support were procured in response, which may have produced some excess.
Army	11181777		X	Electronic Component	Used on Patriot. Highly volatile demand due to varying operating tempo.
Air Force	15142198		X	Hot Section Module, Aircraft Gas Turbine Engine	Underbuy during phase-in; attributed to cost being greater than projected.
Air Force	13430241		X	Exhaust Aircraft Gas Turbine Engine Cone	Unserviceable assets building up because of high condemnation rate; basically converting to consumable item.
Air Force	10657768		X	Servocylinder F-15 Rudder Actuator	Parts shortages primarily due to quality problems (i.e., repair parts unusable). Previously believed contractor semi-starved depot of parts to drive contract repair. Have been inducting for cannibalization.

Table 2.1—Continued

Service	NIIN	Excess	Shortage	Item Description	Case Study Summary
Air Force	12664341		X	Ballistics Computer C-130H, AC-130H, MC-130E (retiring)	Circuit card assembly software problems. Demand is increasing.
Air Force	14707355		X	Receiver-Transmitter, F-15 Fighter Datalink, Multifunctional Information Distribution System	Greatly expanded fielding plan (from 100 to 750) combined with lower-than-projected reliability is producing demands that are higher than projected. No organic repair due to not buying technical data. Period of no demand was due to customers stopping orders in response to very heavy back orders.
Navy	15028101		X	Radar Data Processor, multiple platforms (H53, H60, H1, B22, etc.).	Shifts in repair contracts with gaps between the contracts led to shortages.
Navy	12308172		X	LCAC Marine Propeller	Delay in renewing repair contract (old supplier being bought out contributed to delay).
Navy	12335405		X	Ship Blade Propeller	Low, highly variable demand.

Table 2.2
Excess Case Study Cause Categorizations

		Excess Cause Category					
Service	**NIIN**	**Replaced by Upgraded Item**	**End Item Phased Out/ Fleet Size Reduction**	**Reliability Upgrade of Old Item**	**Program Cancellation/ Plan Change**	**Overbuy of New Item**	**End-of-Life Buy**
Army	014585361					X	
Army	14426926				X	X	
Army	13799894	X					
Air Force	11894176	X					
Air Force	06202517		X				X
Air Force	12348535	X	X				
Navy	200041947			X			
Navy	14353720		X				
Navy	12711063	X					
Air Force	145395245		X				
Air Force	10803407				X		
Air Force	10967677		X				
Navy	13177764						X
Navy	11117804					X	

Table 2.3
Shortage Case Study Cause Categorizations

		Shortage Cause Category							
Service	**NIIN**	**New Item**	**Insufficient Inventory Planned/ Repair Problems**	**Contract Repair Delay**	**Life-Cycle Limit**	**Unable to Repair**	**Parts Short-ages**	**"Not Ready" for Initial Repairs**	**Low Density**
Air Force	145395245						X	X	
Air Force	10803407						X		
Air Force	10967677						X		
Navy	13177764				X				
Navy	11117804				X				X
Army	12844013			X					
Army	11181777								X
Air Force	15142198	X							
Air Force	13430241					X			
Air Force	10657768						X		
Air Force	12664341		X						
Air Force	14707355	X							
Navy	15028101			X					
Navy	12308172			X					X
Navy	12335405								X

Large Disposals of Unserviceable Assets: A Cost of Doing Business Resulting from DLR Phase-Outs and Low Washout Rates

The case studies, supported by consistent information provided during broader process interviews with service supply management organizations, indicate that the disposal of assets is largely a "cost of doing business" when using durable assets that are economically efficient to repair. In other words, these durable assets are simply being disposed of once the DLR is no longer used or used at a much lower rate. These DLRs have typically been repaired many times, potentially upward of 10 or more times over the course of their useful lives, staying in the system once bought. For example, Figure 2.1 shows a DLR for which each asset purchased for and held in DLA DC inventory was used, on average, 104 times over the 10-year period for which we have data.[2] When an end item is replaced or otherwise phased out, the supporting

Figure 2.1
Example High Turnover NIIN

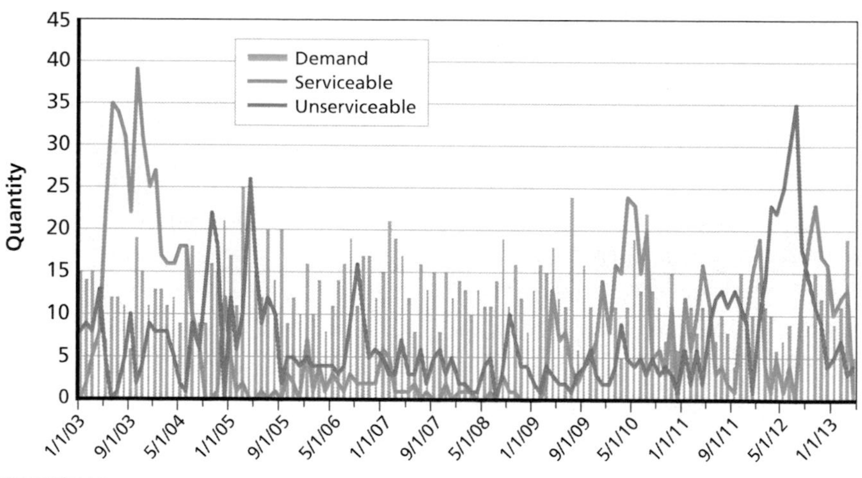

RAND *RR398-2.1*

---

[2] The number 104 is the product of 10.4 inventory turns per year multiplied by 10 years. To determine the number of times each asset has been used overall, we would also need to know the number of this NIIN in retail inventory and the number installed on aircraft.

DLRs are no longer needed. Similarly, when, over the course of an end item's life span, a DLR is replaced by a new version to improve capabilities, to increase readiness by decreasing the replacement rate on the item, or to reduce costs by improving durability and thereby decreasing the replacement rate, the replaced DLR's inventory is no longer needed.[3] Because DLRs are durable assets, their obsolescence should be thought of differently than how obsolescence is typically thought about for inclusion in inventory holding costs. They are much like the durable end items—akin to capital assets in the private sector—they are used on, which are kept until they are no longer needed to conduct DoD missions or until they are replaced by a better end item. For example, the Army has fields full of obsolete end items such as howitzers and trucks at Sierra Army Depot, there are thousands of mothballed aircraft from across the services parked at Davis-Monthan Air Force Base, and the Navy has decommissioned ships anchored or berthed pier-side at a number of locations. In contrast, consumables that become obsolete and are disposed of are purchased but never used. For such consumables, value is never gained, unlike durable assets, such as DLRs, that are used multiple times and then disposed of when they are no longer functionally useful.

Of the 14 case-study NIINs with periods of excess (and all of the severe cases), nine represent this type of general case in which the DLR has been replaced or the end item has been phased out. Another NIIN had a period of excess because of a somewhat large buy meant to last through the end of the life of the end item. One NIIN had a period of excess due to the cancellation of a planned event that would have increased demand (for which more of the DLR was purchased in anticipation). Another DLR is projected to have higher future demands that would eliminate its current excess, and two more DLRs were somewhat overbought in response to one-time demand signals. For the nine phase-out items, all have 0 percent or very close to 0 percent condemna-

---

[3] There are cases, however, in which the carcass is used as the base for the upgraded DLR. The old NIIN is inducted into maintenance, stripped down to the reusable portion, upgraded, and then receipted from maintenance as the new NIIN. In this case, obsolete inventory of the old NIIN is not built up.

tion rates (i.e., when inducted into repair, the depot can almost always repair them), so the assets cannot be gradually phased out through attrition. Rather, as demand declines and then disappears, the assets become excess. The information we obtained from interviews, combined with these data, strongly suggest that the primary reason for DLR excess is item phase-outs combined with very low condemnation rates.

The low condemnation rates for the case-study NIINs are common for DLRs. Figure 2.2 shows the percentage of NIINs (y-axis) with condemnation rates in each of the ranges indicated on the x-axis. The condemnation rate is determined by computing the percentage of receipts from maintenance with a condition code of condemned, or "H." The columns show the percentage of NIINs within the corresponding range of condemnation rates, with the line series being the cumulative percentage for that range or lower. For example, 84 percent of DLRs have had a 0 percent condemnation rate, with 89 percent having a 20 percent condemnation rate or lower. This indicates that,

Figure 2.2
Percentage of NIINs by Condemnation Rate, All Services

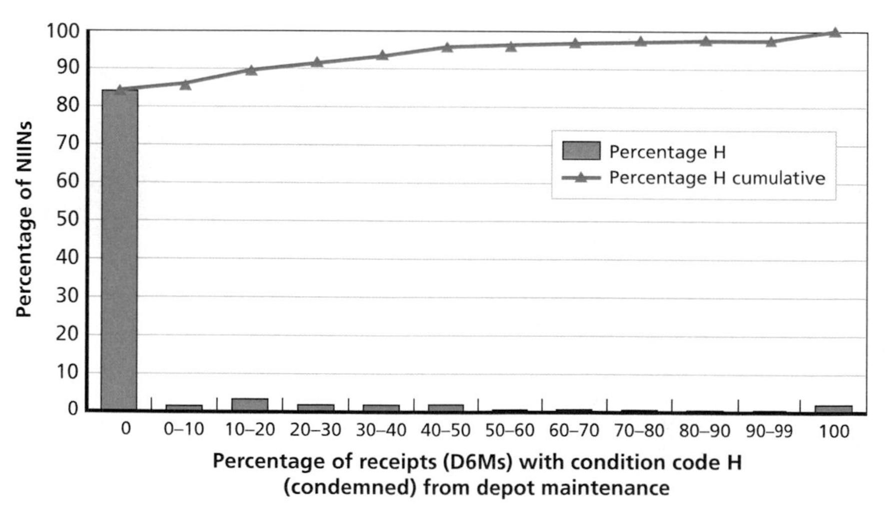

SOURCE: Analysis of DLA MIS Receipts file data.
RAND *RR398-2.2*

for the vast majority of DLRs, as they are phased out, the carcasses will remain in inventory for eventual disposal.

We illustrate this with two representative case study examples. Figure 2.3 shows the monthly serviceable on-hand, unserviceable on-hand, and customer demand quantities for a C-5 brake assembly. This DLR is being phased out and replaced by a new brake assembly with the same functional capability but improved durability, making for less-frequent replacement on the aircraft and thus lower maintenance costs. The old NIIN is being phased out through attrition. That is, it is replaced when condition indicates a need to do so or when aircraft come in for PDM. As it is phased out, demand for the old NIIN has been gradually declining. When worn brake assemblies (old NIIN) are removed from aircraft, they are left in inventory in unserviceable condition, leading to a buildup of such inventory (shown in red in Figure 2.3). Contributing to this buildup is the fact that brake assemblies can almost always be economically repaired; zero washouts or condemnations are recorded in the period shown in Figure 2.3. Thus, it is important to note that the buildup in inventory is not the result of assets purchased to provide spares, but from reducing the installed base, taking

Figure 2.3
DLR Phase-Out Example: NIIN 011894176 Demand and Inventory History

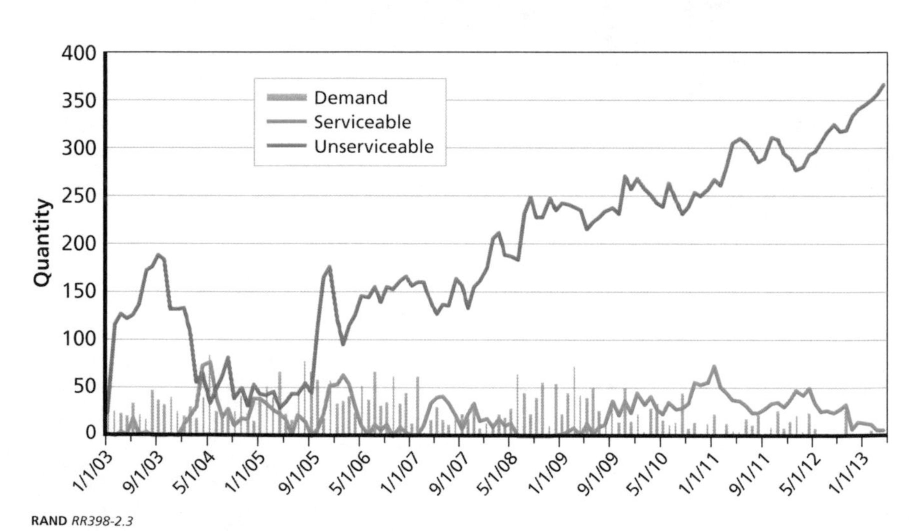

RAND *RR398-2.3*

this NIIN off of aircraft and replacing it with another NIIN. As the installed base is reduced and the brakes are "harvested" from C-5s as they are upgraded, the spares pool of this item becomes bigger than the pool of assets specifically purchased to sustain the fleet. In addition, as the stockpile of carcasses has been built up, serviceable inventory has been kept low so as not to repair into serviceable excess for this item. Work that produces serviceable excess would be wasted repair work. Once fully replaced throughout the fleet, demands will drop to zero, and wholesale inventory will climb further but should be primarily in unserviceable status. Once they are clearly no longer needed, the brake assemblies will be disposed of. However, one could readily argue that this point had already been reached during the period shown in Figure 2.3 and that disposals could have been started with very low risk to readiness. Items are also phased out when DLRs are replaced by new DLRs that offer improved capability, such as a sensor or camera with improved resolution, or increased reliability to improve readiness and reduce costs by reducing unscheduled maintenance actions.

The other main type of DLR phase-out is from the retirement of an end item or a fleet-size reduction. The phase-out of the Marine Corps' H-46E-variant helicopter, for which depot-level maintenance is managed by the Navy, began in FY 2005,[4] with complete retirement of the fleet scheduled by the end of FY 2016 (see Table 2.4). Figure 2.4 shows the demand and inventory history for its transmission. As helicopters are removed from service, demands are declining, leading to a buildup of inventory, again largely unserviceable. The number of assets in the system, planned based on the steady-state period when the fleet was at its peak size, is increasingly higher than needed for the gradually decreasing fleet size, so the excess assets are kept in unserviceable condition. Inventory was built up further because, as helicopters were retired, their transmissions were "harvested" or salvaged from them so that, in turn, parts could be salvaged from the transmissions for use in repair. So as with the C-5 brake assembly, the pool of spare transmissions has become much larger than the pool of assets purchased origi-

---

[4] Gidget Fuentes, "End Nears for CH-46E Sea Knight Helicopter," *Marine Corps Times*, August 23, 2008.

nally to be spares. The large dip in unserviceable transmissions on hand seen in mid-2011 in Figure 2.4 is actually an induction of unserviceable transmissions into maintenance for this salvage operation. This helped alleviate parts shortage problems that had been affecting maintenance. By FY 2017, the inventory of transmissions will be higher yet, with the vast majority of them in unserviceable status for disposal (see Table 2.4). Again, it is clear, though, that the disposal of transmissions could have already started without any risk to readiness. This might have started with the transmissions used to harvest parts from in 2011.

Other Less Common Drivers of Excess DLR Inventory

When plans are put in place to increase the recurring demand for a DLR, the item manager will need to increase the total system inventory to maintain effective support. This needs to be done a lead time in advance of the increase in demand; otherwise, customer support will suffer. Such plans could include an increase in the fleet size of

Figure 2.4
End Item Phase-Out Example: NIIN 014353720 Demand and Inventory History

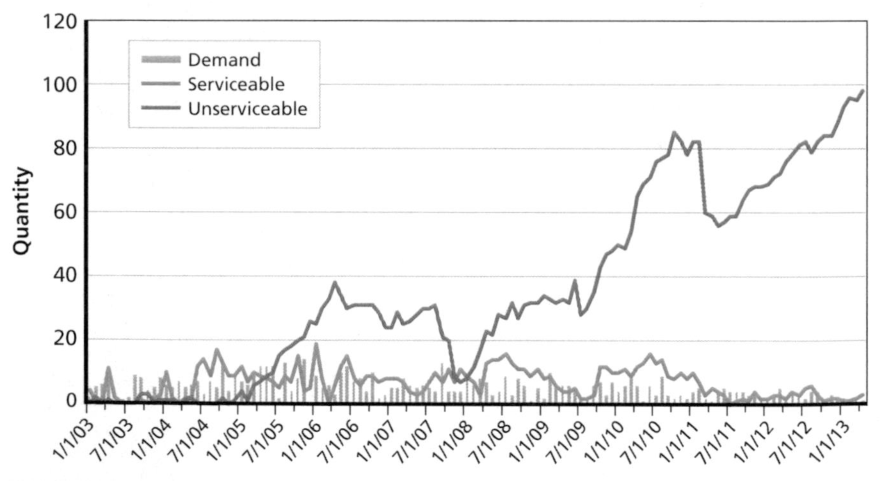

RAND *RR398-2.4*

Table 2.4
Planned H-46E Helicopter Inventory, by FY

Fiscal Year Requirement	FY13	FY14	FY15	FY16	FY17
Total primary aircraft authorized (PAA) start of FY	87	63	39	15	0

SOURCE: NAVSUP Aviation Division, e-mail to the authors, December 14, 2012.

an end item, an enduring change in operating tempo (OPTEMPO), or changes in maintenance plans. If such a plan is cancelled after the additional quantity is ordered and it is too late to cancel the order, these newly purchased assets will become excess, with the excess temporarily serviceable until drawn down because repairs are foregone for a period. Once the excess serviceable assets are drawn down, the excess inventory will be in unserviceable condition. In terms of inventory effect, this is akin to a partial phase-out of a fleet. If the item manager was to wait to see if a plan would be realized, the operational customers would suffer a period of degraded support while waiting for the increase in system assets from procurement.

Somewhat related to this is responding to one-time demand events or signals as if they were representative of a new ongoing level of demand. This can happen if an operational event occurs that increases demand temporarily but the temporary nature of the increase is not clear or communicated to the item manager. This appears to have been the case for one case-study NIIN (014585361, a helicopter engine). An initial demand spike for a new NIIN that was much higher than ongoing demands led to a substantial buy that produced excess. This type of situation can also occur as the result of misinterpretation of demand signals. For one of our case studies, this latter cause led to somewhat elevated inventory, but this elevation was not close to the level of excess inventory seen for the phase-out case studies. Finally, a one-time spike in condemnations from reaching lifetime wear limits led to a large new buy in another case study. However, the excess condition this produced is temporary, as the additional assets will be consumed when the next wave of lifetime repair limits occurs. This case study will be discussed later in the report

An open question is: How long should demand be elevated in order to trigger an increase in system assets? For example, if a three-year operation is planned, it could take additional assets to effectively support the operating units. But, given the durable nature of most DLRs, these assets would become excess after the end of the operation. The buying of assets needed for effective support during this three-year operation could be viewed as an expense related to the operation rather than the generation of inventory excess. An alternative option would be to see if the closed-loop repair cycle time could be temporarily reduced through expedited processes, forward repair, or increased labor, thereby enabling effective support of higher demand through increased maintenance costs instead of an increase in durable assets. The alternative would be poor customer support just when customer support is needed most—during an operation.

Summary of Case-Study Findings for Shortages

Unlike the single broad driver associated with excess DLRs—DLR phase-outs from either end-item phase-outs or being replaced by upgraded DLRs—the causes in our 15 shortage cases were much more distributed, although one cause stands out. The most common factor, arising in four cases, was one or more back-ordered piece parts (which are typically DLA-managed) needed to complete repairs. Additionally, our interviews at the maintenance depots highlighted the issue of parts supportability as by far their greatest constraint on production agility, and maintenance personnel reported high numbers of cases for which they were awaiting parts. At the Air Force Air Logistics Centers and at Navy aviation depot maintenance activities, this issue was the first thing interviewees wanted to discuss because they considered it their most critical barrier to improving support and efficiency. Capacity, repair process flexibility (tooling and labor), and funding were generally not considered to be constraints on repair flexibility in the time frame of this study. However, it is possible that, in the new fiscal environment, funding will become a constraint.

Other root causes identified are as follows:

- In three cases, there were repair contract renewal delays, creating gaps in production. These delays were associated with contracting complications or other factors, including
  - a shift to a new contractor
  - a contractor buyout by another company
  - a contract award delay followed by a delay in product verification.
- In two cases, there were shortages of assets during DLR phase-in/fielding, with different underlying problems affecting asset availability during the phase-in period. These included
  - a late increase in the fielding plan for a DLR, combined with lower-than-anticipated reliability
  - higher DLR price than expected, limiting the initial buy.
- A software problem and insufficient inventory
- The life cycle repair limits were hit for a DLR broadly across the fleet during a concentrated period of time, and the resulting severe spike in condemnations was not anticipated, resulting in a shortage of serviceable assets while waiting for new procurements to arrive.
- A new, unanticipated failure condition that prevented successful repairs and thus led to virtually 100 percent condemnations started occurring for a DLR across the fleet late in the end item's life cycle, resulting in a high level of unanticipated condemnations.
- A planned DLR proved uneconomical to repair, producing a very high condemnation rate that in turn required unanticipated new procurements. This item will be switched to a consumable.
- Two DLRs have low demand with high demand variability, producing challenging forecasting and planning. Two others in this category also exhibited other causes listed above.

Besides impeding customer support, these case studies of shortages correspond to process issues that lengthen repair cycles and affect maintenance agility, increasing inventory requirements to meet customer needs. Thus, they represent process improvement opportunities centered on improving parts supportability, reducing contract lead times, integrated repair planning that considers all resources, and anticipating and planning for knowable shifts in demands and condemnations. Such improvements would reduce costs, as well as improve customer support.

Implications for Measuring Inventory Turns to Monitor Inventory Management Efficiency

As part of its efforts to improve supply chain and inventory management, the Deputy Assistant Secretary of Defense for Supply Chain Integration (DASD(SCI)) is leading an effort in collaboration with the services to establish standard metrics for use by the ASD(L&MR). These metrics will improve oversight and serve as a tool for determining needed supply chain improvements and monitoring improvement initiatives. One such metric is inventory turns, which is a standard metric used in the private sector to evaluate the efficiency of inventory management. For a given level of customer service, the higher the inventory turns, the more efficient the use of inventory (inventory turns and customer service present a trade-off when process capability is constant). The basic calculation of inventory turns is annual sales divided by average on-hand inventory, and the reciprocal is the average amount of time an item is held in inventory before being sold. This ratio puts inventory in relative terms, given that higher levels of demand require greater inventory to meet customer needs. When process performance is unchanged, inventory turns will not change, even as inventory goes up or down in tandem with demand. So using inventory turns enables managers to isolate the effects of the performance of the inventory management process from the effects of changes in demand or sale on inventory levels and costs.

An organization can use inventory turns in two ways: (1) to benchmark relative inventory performance against other organizations with similar demand environments to determine whether there is opportunity to implement better practices and (2) to assess whether it is improving inventory management processes over time. Note that, for benchmarking, determining the right comparisons is critical; inventory turns vary by orders of magnitude for the top performers in different industry segments based on factors such as lead times, demand levels, and demand variability. In practice, establishing the right benchmark for some organizations can be difficult. This is particularly so for DoD at large and for some segments of DoD supply, such as DLRs, that are similar to only a small number of private sector organizations and are not fully analogous even to these (e.g., many weapon system DLRs are more complex, more expensive, lower in demand, and have a higher demand variability than most such items in the private sector). But even if external benchmarking cannot be done well, using turns to benchmark internally or monitor progress over time can be valuable for monitoring how well an organization is progressing in its efforts to improve its planning and management of inventory. It is important, though, to use and measure turns in ways that ensure the metric aligns well with this purpose; the metrics should align with process performance, moving in tandem with it to provide effective feedback.

Thus, the findings in this chapter regarding the causes of excess inventory and disposals have significant implications for how inventory turns should be looked at for DLRs. Including all DLRs in the calculation of inventory turns would not provide a good picture of process performance because the total population consists of several subpopulations with very different but explainable levels of inventory turns. In particular, total DLR inventory includes a substantial inventory of items that are being or recently have been phased out, producing very low turns for these durable assets. Combining the different categories does not produce a meaningful picture for assessing inventory and overall supply chain management effectiveness.

Accordingly, we developed an approach for categorizing DLRs into different life-cycle phases, as described in detail in Appendix E, so that inventory turns can be measured by DLR category, providing an accurate picture of process performance. The first category, Phase In, represents the period during which the DLR is being introduced into an existing end-item fleet or the fleet is in its initial fielding stage. Steady State is the period during which demand is relatively stable, occurring when a fleet size is roughly stable and before a DLR is replaced by an upgrade, if ever. Phase-out represents the period during

Figure 2.5
Example Item Categorizations by Life-Cycle Phase

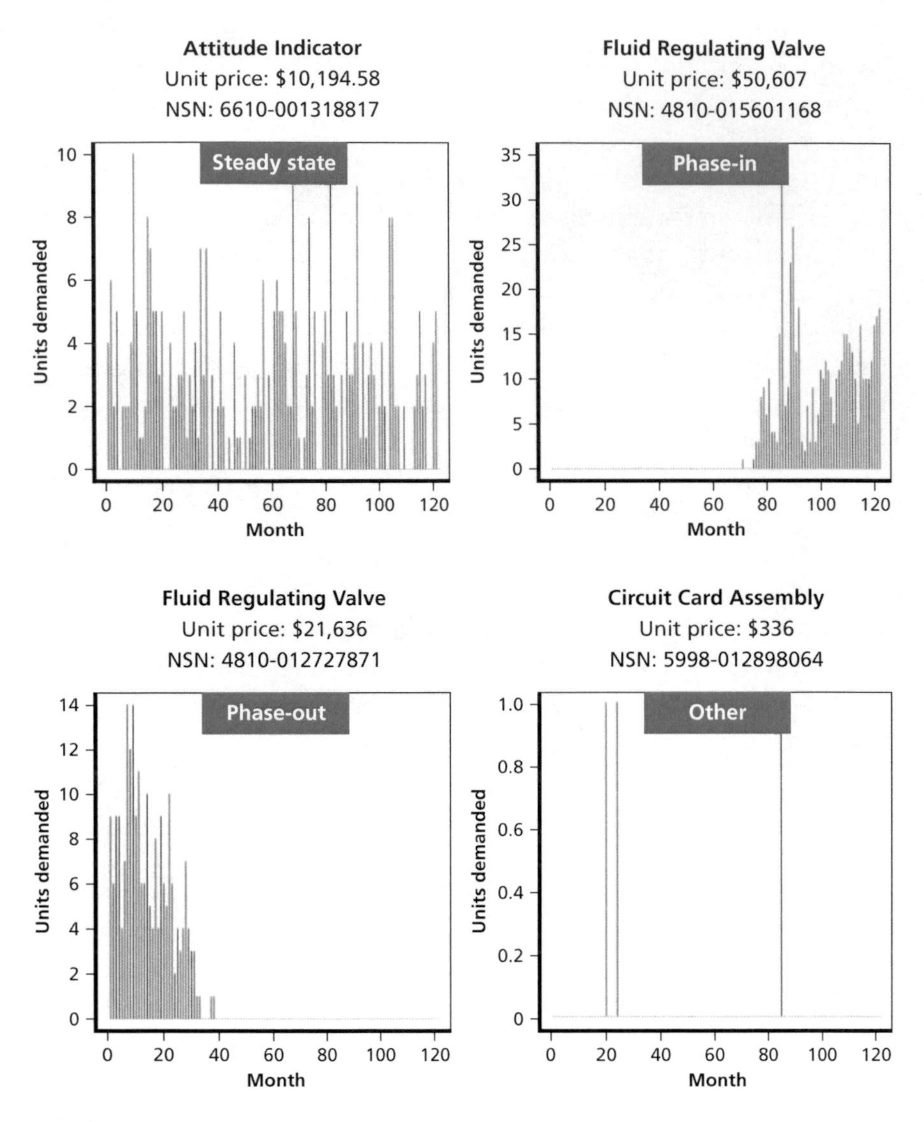

RAND *RR398-2.5*

which demand declines as a DLR or the end-item fleet is phased out. We categorized items with no demands (which may have been phased out before the start point of the data available) and items with very low, sporadic demands into the category of Other. Figure 2.5 shows examples of DLRs that demonstrate the four types of patterns by depicting monthly demands over 10 years. The item representing the Other category had just three months with one demand each in 10 years.

After categorizing all DLRs, we computed turns by category and service to show their differences. The data were limited to DLR inventory held in DLA DCs and issues from the same, as this was the extent of the available historical inventory and transaction data.[5] This analysis reveals that these "wholesale" turns for steady-state items range from 0.6 to 1.4 among the services, as shown in Figure 2.6. In other words, each DLR held in inventory at DLA DCs goes through an entire closed-loop use cycle every 9–19 months, depending on the service. Phase-out items naturally have much lower turns, and items in the Other category also have very low turns. Notably, while most demand is in the Steady State category—two-thirds to three-fourths across the services (see Figure 2.7)—the Phase-Out and Other categories include about half of the inventory (Figure 2.8), and the vast majority of items—80–90 percent—are in the Other category (Figure 2.9). We also see in Figure 2.6 that phase-in items have relatively high turns, indicative of two situations: (1) a greater propensity of customer support difficulties than excess inventory for items being phased in and (2) items replacing other items as upgrades that are being phased in gradually over time. In the latter case, inventory on the shelf can be limited in accordance with the phase-in plan. In other words, the fact that turns are higher for phase-in items is not indicative of better management. Rather, it indicates too little inventory for some items and a deliberate, gradual phase-in strategy for other items.

This nuanced perspective in terms of measuring performance is important. Process improvements in DLR supply chain management will primarily affect steady-state management. The point in a DLR's

---

[5] As issues from DLA DCs, these may not be the same as sales, since some would represent repositioning within a service Working Capital Fund from "wholesale" inventory at a DLA DC to a retail inventory point within the same Working Capital Fund, representing a retail inventory replenishment. This turns calculation will not produce the same result as total system turns (including these retail stocks) computed using sales from the Working Capital Fund.

Figure 2.6
Inventory Turns by NIIN Category

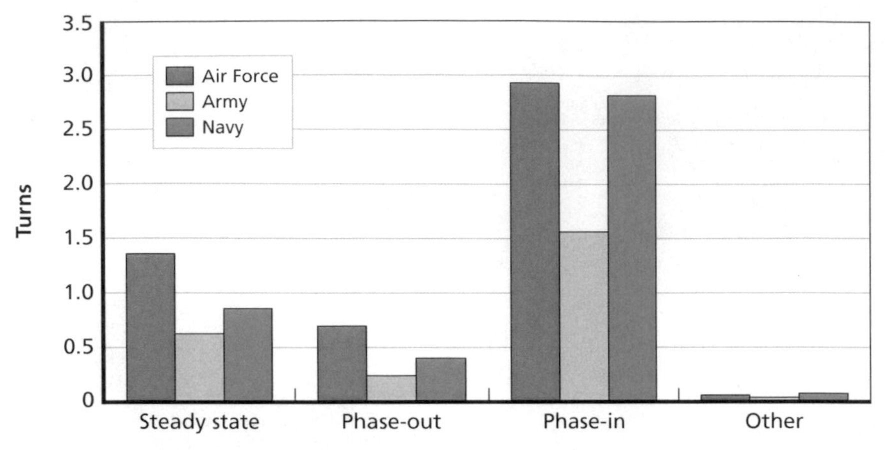

RAND *RR398-2.6*

life cycle at which it has the maximum steady demand will determine the maximum amount of inventory needed in the system for customer support (assuming a steady closed-loop cycle time from the removal of a DLR from an end item to reissue in serviceable condition for use).

Figure 2.7
Percentage of DLR Demand by NIIN Category

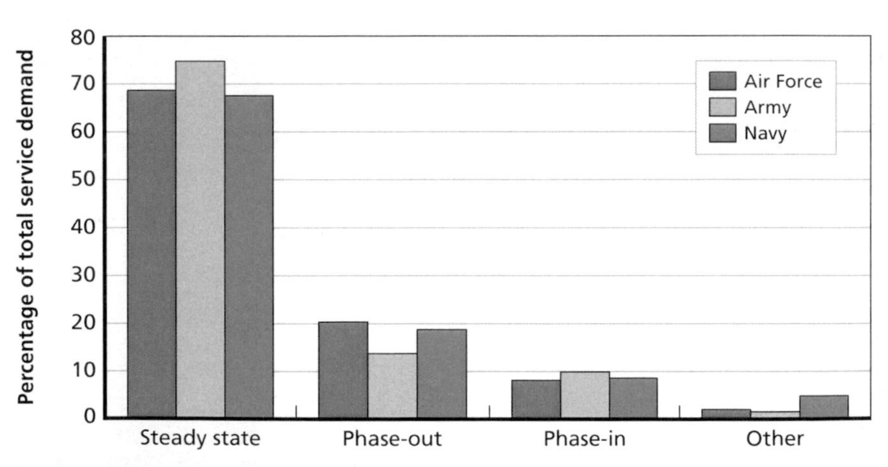

RAND *RR398-2.7*

Figure 2.8
Percentage of DLR Inventory Value Held in DLA DCs by NIIN Category

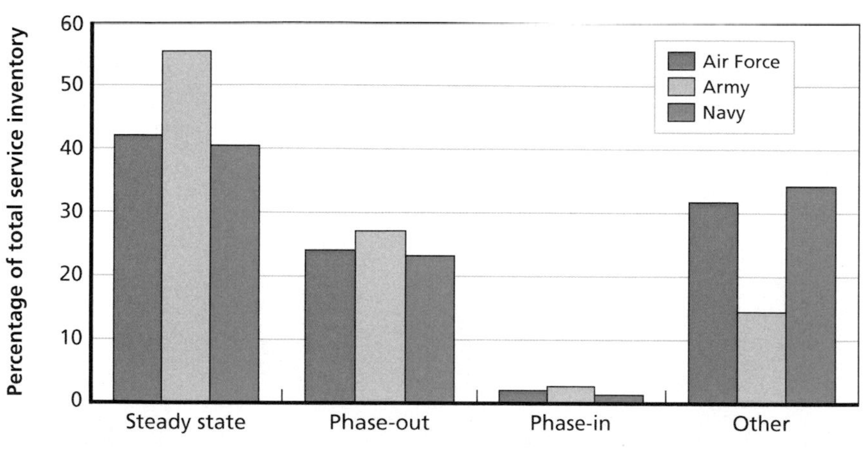

RAND *RR398-2.8*

For most items—those with no condemnations or very low condemnation rates—this period will produce the amount of inventory that will be in the system until disposals begin. Turns measured outside this

Figure 2.9
Percentage of NIINs by NIIN Category

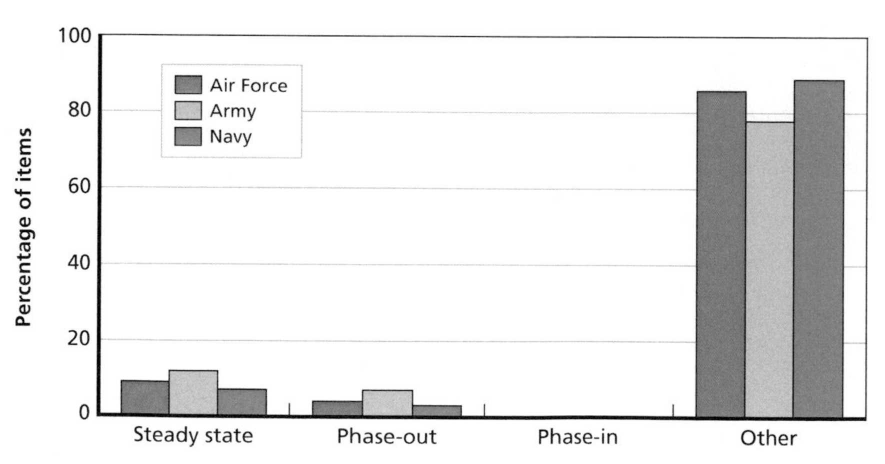

RAND *RR398-2.9*

maximum steady-state period will not give an accurate picture of the performance of the inventory management process.

Inventory turns is a valuable management metric to help monitor and improve DLR supply chain management efficiency, and the DASD(SCI) and ASD(L&MR) should continue efforts to make it a standard DoD metric. But in doing so, this metric should be limited to those DLRs in the steady-state phase of their life cycles to ensure it provides meaningful and actionable information.

CHAPTER THREE

Parts Supportability/Taking a Total-Cost Perspective

Parts supportability issues clearly impede customer support, and this effect of the problem gets significant attention within the services and their discussions with DLA. Our visits to maintenance depots across the services also revealed a number of common practices in the depots for dealing with or stemming from shortages for production that increase maintenance costs. Most of these increase depot labor costs, which is an effect of the problem that receives less attention. These practices or workarounds include

- cannibalization of work-in-process assets and the induction of carcasses solely for the salvage of parts
- local fabrication in lots of one or small batches
- rework when production continues despite the absence of a part (i.e., leaving one part out of a completed DLR, requiring some teardown and rebuild later)
- moving work-in-process to an awaiting parts storage area when a part shortage occurs
- line shutdown when the absence of parts prevents production
- the redeployment of labor to areas in which the personnel are less effectively trained, impeding productivity
- adding extra capacity to handle greater variability stemming from parts shortages
- local purchases, such as those made through credit cards.

Data systems do not capture these events, nor do they capture the amount of work involved in each. Thus, quantifying the associated costs is not possible. However, at several depots, this issue is of great concern and their local data snapshots and interviews with their personnel indicate that substantial amounts of work are affected by parts shortages, suggesting that the costs could be substantial.

During our interviews, this issue came across as the most severe at ALCs and Navy aviation depots. And our interviews, in conjunction with observations during visits, suggested this may stem from a few factors. Navy shipyards have substantial machining/local fabrication capabilities and capacity to support ship availabilities. These capabilities are also used to fabricate parts. These capabilities may be sufficient enough and such a standard part of doing business that they mask the impact of parts shortages. On the Army side, there are three factors that could be lessening parts supportability issues or mitigating their effects. As we will discuss later, the Army tends to employ push production, which, while it has some disadvantages, could facilitate parts planning by providing longer production planning lead times and parts ordering horizons. Second, the Army DLRs tend have higher demand levels due to greater fleet sizes, and this tends to improve parts supportability.[1] Additionally, the Army has maintained its own retail-managed and owned inventory at its depots, unlike the Air Force and Navy. As a result of the Base Realignment and Closure (BRAC) legislation of 2005, the Army turned over labor of these warehouses to DLA but not these other functions. It is possible the Army is stocking more richly at the depots, implicitly recognizing the productivity costs of parts shortages—especially with a push system—and with such parts representing a relatively small investment compared to DLR investment.

The services are cognizant of this issue and place substantial emphasis on improving parts supportability, with what appears to be the greatest emphasis by the Air Force. With the transition of parts management for depot maintenance to DLA as a result of the 2005 BRAC legislation, there is an increased need for effective cross-organizational coordination. Following perceptions of early struggles, the services are taking an increasingly proactive role in improvement,

---

[1] The average monthly demand for steady items, as defined in Chapter Two, was 14 units for the Army versus three for the Air Force and two for the Navy.

both by focusing on what they can do to improve the situation and by being demanding customers.

Demand History Adjustments

One area of emphasis in the Air Force, starting in 2012, has been the increased use of demand history adjustments (DHAs). When a part is not available and the depot does not submit an order for the part, executing a workaround instead, a demand record is not created. This impacts the demand history that DLA relies on to forecast future demands and thus plan inventory. So the workarounds that maintenance executes can thus impede the improvement of support to these operations. By finding ways to work around the problem and not reporting their needs or the workarounds, maintenance activities mask the extent of the problem and do not provide DLA with the information it needs to better plan parts support. DHAs allow the depot to submit a demand record that gets integrated into an item's demand history and thus into DLA demand planning. From the viewpoint of a DLA demand planner, the DHAs are transparent, being treated as any other demand. The ALCs indicated in interviews that they started emphasizing the submission of DHAs in 2012. DLA data do indicate that the Air Force has indeed been increasing its use of DHAs since 2009, when it actually had a decline in use compared with 2008. The same data indicate that the Navy is the biggest user of DHAs, although with a recent decline. The Army and the Marine Corps make limited use of DHAs. Figure 3.1 shows these trends.

Collaborative Planning

The second area of Air Force improvement attention focuses on improving collaboration through improvements in demand data exchange (DDE). When the services have plans that will lead to demands diverging from historical trends, they can submit projected demands in the form of DDE data. Recognizing the need to be proactive in this regard,

Figure 3.1
Demand History

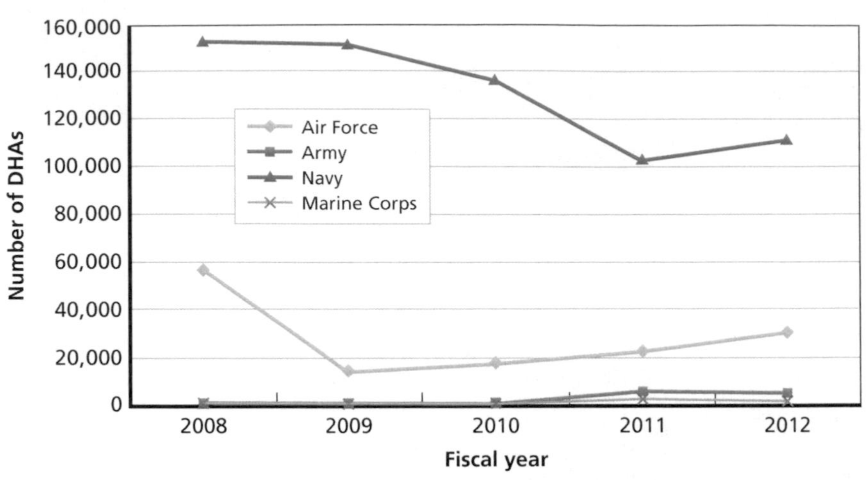

SOURCE: Based on data query of DHAs by DoD Activity Address Code conducted by the DLA Office of Operations Research and Resource Analysis (DORRA).
RAND *RR398-3.1*

the Air Force developed standardized DDE generation and submission practices and codified them in a new Air Force instruction in 2012.[2] This included the establishment of a Planning for DLA-Managed Consumables (PDMC) Flight to centrally coordinate the process across the Air Force. The PDMC Flight gathers information on projected changes in demand from the Air Force's maintenance and supply communities, serving as a single point of contact to coordinate and submit DDEs to the DLA. One related area of emphasis among some of the ALCs pertains to DLRs awaiting parts. When a consumable needed to repair a DLR is short, it can lead to a decision to shut down the induction and production of the DLR altogether so as not to develop DLRs that are partially repaired but awaiting parts (AWP). This, in turn, suppresses demand for the other consumables used on the DLR. When the shortage item for the DLR's production is replenished, it may be followed

---

2 Headquarters, Air Force Materiel Command, *Material Management: Planning for DLA-Managed Consumables (PDMC)*, Air Force Materiel Command Instruction 23-205, April 26, 2012.

by a shortage of one of the other items, since the apparent demand for these items would have dried up, affecting DLA's forecasts for them. To avoid this problem, the depots have started submitting DDEs to indicate that demand for these temporarily non-ordered NIINs will restart.

The Navy and Army are taking a different approach, enabled by their adoption of SAP for their enterprise resource planning (ERP) systems in conjunction with DLA's adoption of SAP. They are electronically tying their ERPs to DLA's to directly transmit parts needs associated with their depot production plans to DLA. The Navy went live with this, with a launch at FRC Southwest at Naval Air Station North Island, California, in June 2013. It will be phased in to the other FRCs one at a time, with the second launch scheduled for FRC Southeast at Naval Air Station Jacksonville.[3]

The Manufacturing Resources Planning (MRPII) module within SAP produces what is called a *gross demand plan* detailing the parts needed to support production. The change made by the Navy and DLA is for the Navy's ERP to pass the gross demand plan directly to DLA's ERP for direct use within the demand planning process. Thus, when fully phased in, the Navy production plan will directly produce the portion of the DLA demand plan for DLA's Navy depot maintenance customers, with a two-year production forecast provided. This will be combined with historical demands from and collaboration with non-depot customers to create the overall demand plan.[4] At the time of this writing, it is too early to evaluate the impact. The Army process will be similar, with the additional incorporation of a method to subtract or "net out" on-hand, Army-owned stocks in its depots.

---

[3] Office of the Commander, Fleet Readiness Center, Naval Air Systems Command, correspondence with the authors, August 15, 2013.

[4] Interviews with DLA HQ personnel; Defense Logistics Agency, "IMSP [Inventory Management and Stock Positioning] Spiral 2 Gross Demand Plan Process Flow Overview," briefing, May 15, 2013.

Accounting for the Maintenance Costs from Parts Shortages

Beyond these efforts to improve information flow to DLA for improved planning, the consequences of parts shortages suggest another path DoD could take toward improvement. Service-level goals for DLA safety stock are based on service-level agreements with the services and loosely based on readiness needs and historical levels of support performance. In the private sector, when there is a stockout, there is risk of a lost sale—the loss of the prospective benefit of having an item in stock. In the military, the benefit of having an item in stock is readiness, which is difficult to turn into "revenue" value to trade off against inventory when considering stockout costs versus inventory costs. However, per our earlier discussion, there is a monetary value component to stockouts in support of depot maintenance—additional maintenance costs. These could be substantial, but without more in-depth examination, they are unknown. With these costs completely unknown, they are not considered in planning inventory levels. In contrast, if there were reasonable estimates of stockout costs, they could be considered in tandem with safety stock investment costs to determine the lowest total cost level of safety stock, incorporating both inventory costs and maintenance costs. It is possible that more safety stock could lower DoD costs if the additional safety stock lowered maintenance costs more than it increased inventory holding costs. It is also possible that this would suggest a lower safety stock level but then one could consider whether or not to keep the current level based on readiness needs. In short, the optimal service level might be the higher of the readiness-based safety stock level and the total cost minimizing level.

Expanding on the concept of integrating stockout costs into safety stock planning, there is also the consideration that parts do not affect maintenance in isolation. Rather, there are part dependencies. It is the work-order or maintenance-line perspective that maintenance cares about—not just item-level materiel availability. This has two implications. The first is that metrics that measure effects on maintenance could be important to track. Second, not all part stockouts will generate the same cost. It depends on the workarounds executed to make up for the shortage. This might have a readiness linkage or dependency. For example, for readiness needs, if a DLR is stocked out, it might be important for the depot to find a way to repair carcasses of the DLR no matter what. In contrast, urgency could be much lower for a DLR with serviceable stock on the shelf. The cost also depends on the overall impact on the maintenance shop or production line. If a line only produces a limited number of DLRs, the inability to produce any one DLR due to a part shortage would have a higher cost impact than it would on a line with significant flexibility that could turn to other DLRs without much or any impact on productivity. In an extreme case, a line that could only produce one DLR would be at risk of high maintenance costs in the face of parts shortages, as the entire line would be idle and the labor might not be redeployable and productively used elsewhere. One good example of this is the landing gear shop at Ogden Air Force Base. It has two production lines, with one set of tools and fixtures for the three heaviest landing gears and another set for the rest. A majority of the work on the heavy line is for C-5 landing gears. So when the shop is missing parts that prevent C-5 landing gear repair, the utilization of the line plummets, increasing costs due to idle capacity. As this illustrates, the costs of all parts shortages are not equal, potentially signaling the need for safety stock prioritization based on costs, which would naturally occur if stockout costs were considered variable for different groups of DLRs. This would move DLA, and DoD overall, toward a DLR-level parts supportability performance perspective rather than an item or piece part–level perspective.

Parts Supportability Metrics and Performance-Based Agreements

A review of performance-based agreements (PBAs) with DLA across the services reveals limited focus on the maintenance or DLR-level perspective in the PBAs and significant differences in PBAs across the services. Table 3.1 shows the metrics in each PBA. The Air Force, Marine Corps, and Navy do share a relatively common core of item-level supply chain performance metrics in their PBAs that are focused on response

Table 3.1
Service-DLA PBA Metrics by Service

Metric	Definition	Air Force	Army	Marine Corps	NA
AWP	Items for which DLA is the only SOS impacting the Air Force–managed end item NSN	X			
ORT	Percentage of ALC maintenance documents filled within two days (DLA APR, not PBA)	X			
CWT/LRT/TDD percentage	Air Force has separate goals for ALCs	X	TDD	X	X
Material availability	1 – number of unfilled orders established/number of orders			X	X
Unfilled orders	Unfilled orders currently on hand			X	X
Demand planning accuracy (DPA)	$DPA = \left[1 - \frac{\lvert Fcst - Hist \rvert}{Max(Hist, Fcst)}\right] \bullet 100$	X		X	X
POF	Number of perfect orders/number of total closed out orders	X		X	X
Price change	In development	X		X	X
Inventory turns	12-month sales/issues/average inventory on hand	X		X	X
Eng. support quality	Percentage of the executable requests returned to DLA	X			
Eng. support timeliness	Percentage of work in process that has not exceeded the due date	X			
DLR attainment to plan (ATP)	Using OTD: percentage of time vendors deliver in accordance with the terms of the contract	X			

Table 3.1—Continued

Metric	Definition	Air Force	Army	Marine Corps	NA
Not mission capable (MICAP) AWP hours	Sum of hours a customer waits for a part that grounds an end item	X			
Inventory accuracy	Accountable inventory records compared to physical inventory				X
FDD weight fill rate	Percentage of weight filled from the designated FDD		X		
Strategic fill rate	Percentage of issues from the supporting SDP and FDD		X		

SOURCES: Headquarters, Defense Logistics Agency, Air Force Military Service Support Team, and Headquarters Air Force, ILCM Policy Division, "Logistics, Installations & Mission Support United States Air Force (AF/A4/7) and Defense Logistics Agency (DLA) Performance Based Agreement," Version 3.0, September 20, 2010; Headquarters, Air Force and Headquarters, AFMC/LG, "AF/AFMC/DLA Performance Based Agreement Addendum," March 14, 2012; Performance Based Agreement Between Director, Defense Logistics Agency and Chief of Naval Operations and Commander, Naval Supply Systems Command signed February 16, 2011, March 25, 2011, and March 8, 2011, respectively; Headquarters, United States Army G-4 and Headquarters, Defense Logistics Agency, Performance Based Agreement, May 12, 2008; Marine Corps Team, Military Service Support Division, and Defense Logistics Agency, Performance Based Agreement Between the Defense Logistics Agency and Headquarters, U.S. Marine Corps, Version 3.0, July 8, 2010.

NOTE: CWT = customer wait time, LRT = logistics response time, TDD = time definite delivery, Fcst = Forecast, Hist = actual demand history, POF = perfect order fulfillment, FDD = forward distribution depot, SDP = strategic distribution platform.

time and item availability. However, the Air Force adds several metrics that provide a maintenance and DLR perspective. AWP is the most directly relevant metric, recording the number of maintenance items for which DLA is the only source of supply impacting the Air Force's ability to complete a repair. Order response time (ORT) is focused on ALC maintenance, measuring response time for parts to support ALC repairs at the item level. DLR attainment to plan records the percentage of time vendors deliver new buys of DLRs on time. The MICAP hours metric records how DLA is affecting overall Air Force readiness; it can be impacted by shortages of parts for DLR repair or shortages for the replacement of items directly on aircraft. The Air Force's overall PBA with DLA is also unique in that it requires timeliness from the Air Force in providing engineering support to the DLA contracting process, making lead-time management of DLA parts more of a two-way street. In this vein, the Air Force–DLA PBA includes demand-plan accuracy, which ideally would be decomposed to reflect the effectiveness of Air Force information-providing processes regarding demand changes and DLA response to this information.

It should also be noted that the Army has few metrics in its overall PBA with DLA. The Army-DLA PBA lays out a suite of metrics that could be used by individual Army organizations, such as an individual depot, to develop PBAs with DLA. This leaves few metrics as part of the overall Army-DLA PBA.

In addition to using a broader suite of metrics, the Air Force has worked with DLA to ensure the metrics get high-level attention. As a result, the DLA director includes metrics that are important to the Air Force in DLA's monthly Agency Performance Review (APR) meetings, which are attended by the director's direct reports and the remainder of the DLA senior leadership team. Several Air Force maintenance–related metrics were in the March 2013 APR, including ORT and AWP, as seen in Figure 3.2, which shows two charts from the APR briefing and meeting. The left-hand graph shows the actual ORT or percentage of ALC maintenance requests filled within two days by month versus the goal. The right-hand graph shows the number of ALC AWP work orders by month versus the goal. In addition, the total number of back

Figure 3.2
DLA Aviation Industrial Performance

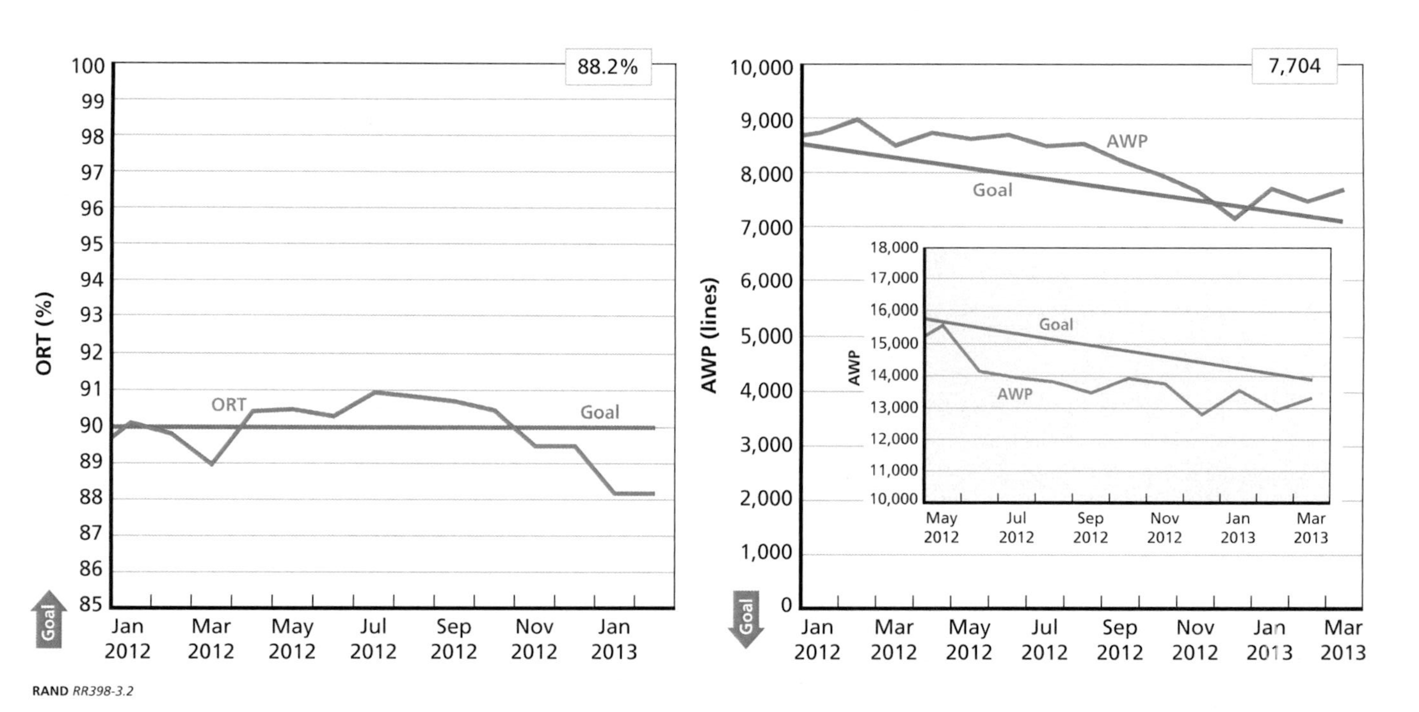

RAND *RR398-3.2*

orders against the AWP lines is shown in the small graph overlaying the larger AWP graph.

Example Improvement

While there is not systematic evidence, an example suggests that these types of activities together can improve parts supportability. The landing gear shop at Ogden ALC has undertaken a number of initiatives that, in concert, have led to improved parts supportability. First, rather than relying on DLA to provide parts through inventory as needed, it has developed improved planning tools and conducts checks of DLA inventories as part of the production planning process. This enables it to communicate to DLA which items are looming problems as opposed to waiting for AWP status. Second, it has focused on ensuring the rigorous use of DHAs. Third, it has focused on using DDEs to reflect demands for NIINs that have been temporarily suppressed due to shortages of other parts. Over a two-year period, these efforts resulted in a dramatic reduction in the number of piece parts causing AWP conditions, as shown in Figure 3.3. There is a belief that continued improvement will be difficult, as many of the remaining problem national stock numbers (NSNs) have long lead times.

Figure 3.3
Ogden ALC DLA Problematic NSN Timeline

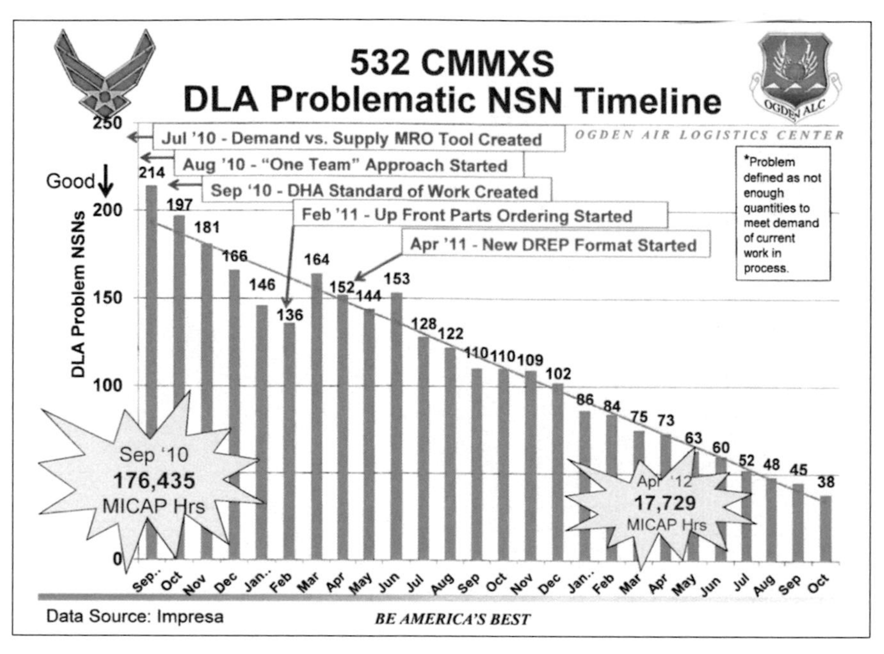

SOURCE: Chart provided by the 532 Commodities Maintenance Squadron.
RAND *RR398-C.3*

CHAPTER FOUR

Adoption of Pull Production

Aligning supply with demand is a hallmark of pull production. We aimed to determine the degree to which each service employs a pull production system and whether there might be value in shifts in degree of push versus pull for any of the services.

Production System Profiles

The Air Force system can be described as very close to a complete pull system. Each day, its DLR production planning tool, called EXPRESS (Execution and Prioritization of Repair Support System), looks at current inventory levels, outstanding demands and their priorities (including whether the demands are associated with mission capability), and depot flow times to determine the prioritization of repair induction for that day. Production is prioritized first to outstanding MICAP demands with no on-hand serviceable inventory and last to bringing inventory levels up to the maximum serviceable inventory target. Therefore, the system is very much tied to daily demand and will not repair into serviceable excess. The result is a relatively lean system in terms of serviceable inventory, resulting in relatively high back-order rates. But prioritization appears to minimize the readiness impact, with EXPRESS also prioritizing distribution of serviceable items to the customers in greatest need. There is significant excess of some items, but this is generally due to phase-outs of DLRs, as discussed earlier.

Within the Air Force, there has been some debate about whether daily planning introduces excessive turbulence into maintenance processes, degrading productivity and increasing total cost. However, in our site visits and interviews, we learned that the different ALCs and shops within them, including some that technically employ daily induction planning, are employing different practices to apply limited level-loading for maintenance efficiency while preserving the pull production paradigm. This will be discussed in more depth later.

The Navy employs a combination of pull production and what we term *modified pull production*. Over half of the Navy's DLR repair is done at intermediate-level maintenance.[1] These activities operate on a pull system with one-piece flow. When DLR fails and is removed from an aircraft or ship, if the needed repair action is within the capability of intermediate maintenance, it is sent there for repair to be returned to the customer if there is no serviceable stock and to inventory otherwise. If the DLR cannot be repaired at the intermediate level, it is sent to storage for induction into depot maintenance.

NAVSUP workloads depot-level or level-III maintenance activities. At the start of the FY, NAVSUP gives them six-month induction plans. At the start of the third quarter, WSS Mechanicsburg, which manages supply for ships, provides a second six-month plan to the shipyards. WSS Philadelphia, which manages supply for aviation, provides three-month plans at the start of the third and fourth quarters. The depot maintenance activities determine the detailed production schedules within these periods, balancing production efficiency and customer needs if shortages of an item are communicated to them during the period. As will be shown, this semiannual/quarterly workloading appears to produce a serviceable/unserviceable profile similar to that of the Air Force. The result is that this frequency of planning appears sufficient to produce results close to a pull system. The frequency is sufficient to not severely overproduce or underproduce, and changes can be made within the periods by exception. Shipyards also integrate NAVSUP workloading with direct repair and return to ships in dock for availabilities, with the NAVSUP workload being a small portion of their volume.

---

[1] Based upon analysis of data provided by NAVSUP, which provided a database with one year of repairs through a query of its ERP.

The Army process appears closer to a push system in design, with annual workloading provided to the depots and changes by exception during the year. This tends to limit changes to the most severe cases and is a less reliable process for making all needed changes. For example, Figure 4.1 shows scheduled versus actual production by Army depot. A complete push system with no constraints will show equal quantities for each NIIN when comparing actual production with the year's planned production. While some difference between the two can be seen, as indicated by points off the 45 degree lines in the graphs in Figure 4.1, it is relatively limited in degree, and many items are produced exactly as planned in the year's initial workload plan. Reduced production sometimes results from resource shortages, such as insufficient carcasses, rather than from changes in plans.

A modified pull system requires a new workload plan on a periodic basis—production will stop otherwise, and a complete pull system will only produce on demand. In contrast, a push system requires an exception decision be made for every NIIN for which demand has changed significantly—it will continue as planned unless altered. A modified pull system forces a decision on every NIIN at the frequency of the planning horizon. This is the fundamental difference between push and pull: whether changes to the plan are made by exception or changes are built into the process on a systematic, more frequent basis. As a result, we observe more Army NIINs with overproduction than we do for the Air Force and Navy, as well as some with underproduction. This is relatively limited, though, in times of stable demand. The effect becomes greater in times of shifting or highly variable demand.

To show the difference at the aggregate level, Figure 4.2 shows the percentage of DLR inventory in DLA DCs by service that is serviceable versus unserviceable. Consistent with the process design, the Air Force has the lowest serviceable asset percentage, the Navy's percentage is a little higher, and the Army has the highest percentage. Higher percentages could either be good or bad. They potentially—but not necessarily—provide lower back-order rates, but they could also represent increased risk of having repaired into long supply.

Table 4.1 shows these percentages by service and life-cycle category, with the Air Force and Navy patterns being similar. Even in

Figure 4.1
Indication of Army Push: FY 2012 Repairs Versus Plans by Army Depot for Individual NIINs

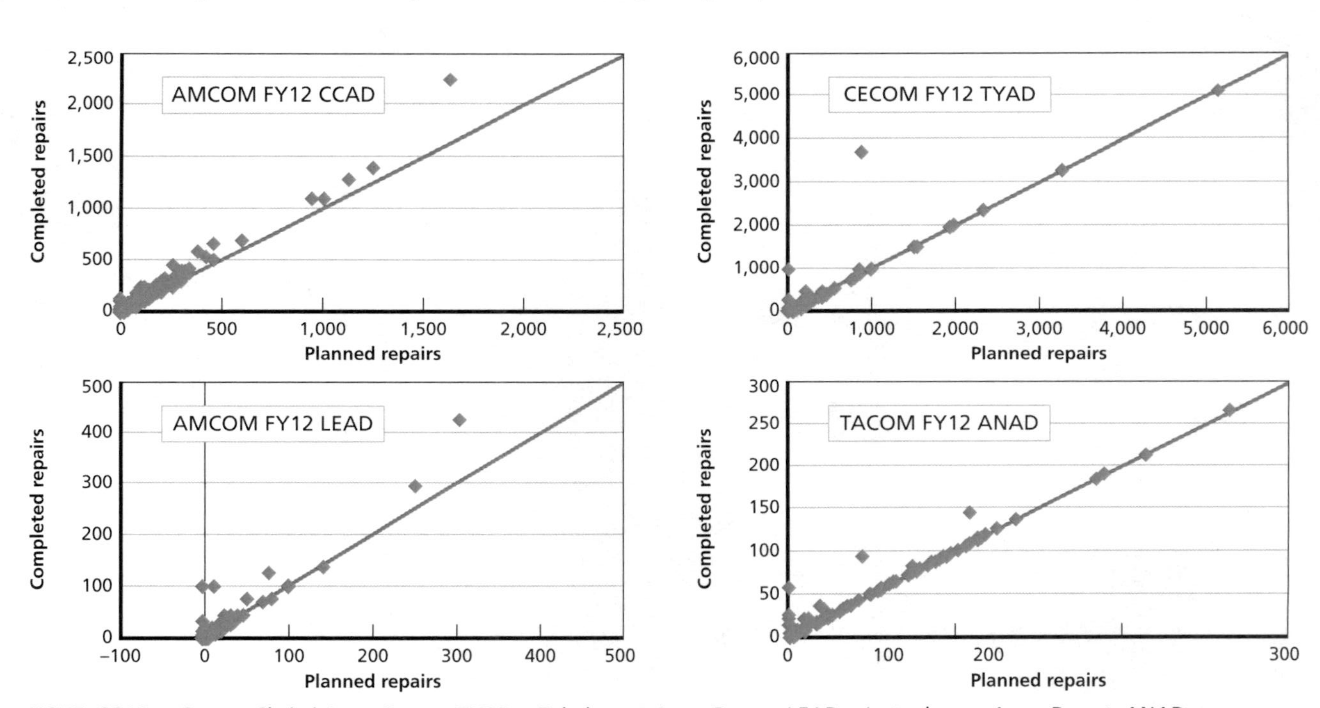

NOTE: CCAD = Corpus Christi Army Depot, TYAD = Tobyhanna Army Depot, LEAD = Letterkenny Army Depot, ANAD = Anniston Army Depot.

RAND *RR398-4.1*

Figure 4.2
Serviceable and Unserviceable Inventory Percentages by Service

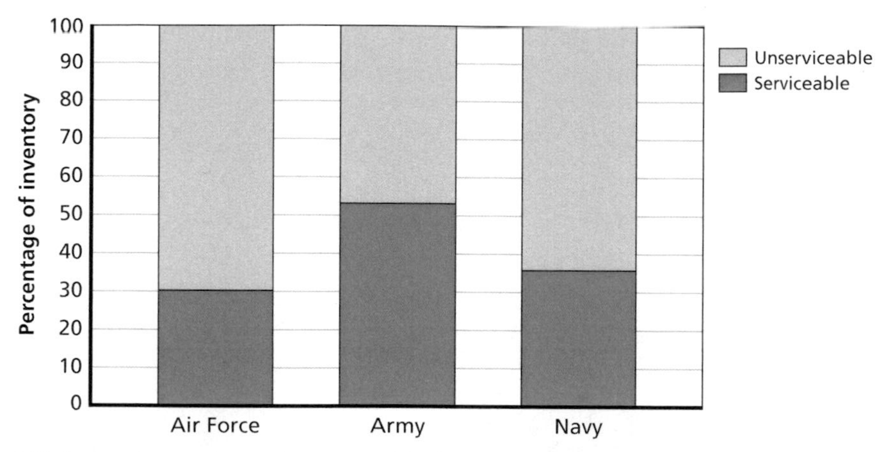

NOTE: Based on analysis of DLA QBO file data, 2011 averages.
RAND *RR398-4.2*

Table 4.1
Serviceable and Unserviceable Percentages by Service and DLR Life-Cycle Category

Service	Steady State	Phase-In	Phase-Out	Other
Air Force				
Serviceable Inventory	32%	68%	21%	44%
Unserviceable Inventory	68%	32%	79%	56%
Army				
Serviceable Inventory	60%	72%	47%	42%
Unserviceable Inventory	40%	28%	53%	58%
Navy				
Serviceable Inventory	35%	51%	26%	44%
Unserviceable Inventory	65%	49%	74%	56%

SOURCE: DLA QBO file data, 2003–2012 averages.

the Steady State category, both tend to keep a majority of assets in unserviceable condition, keeping repair relatively aligned with demand on wholesale inventory (often even after a demand to replenish retail inventory), although the high unserviceable percentage for steady-state items may indicate some room for improvement in overall inventory or the potential to get slightly ahead of demand to provide better customer support. In some cases, though, the high percentage of unserviceable assets for a NIIN is due to parts constraints on production. In the Phase-Out category, in which it becomes advantageous to reduce the relative percentage of serviceable assets, the Air Force and Navy have lower percentages of these assets in serviceable condition than the Army, carrying over from the steady-state phase. The higher Army percentage of serviceable assets for phase-out items could also be a function of the slower response to decreases in demand that results from a more push-like system as items begin to be phased out. For all three services, the percentage of phase-in assets in serviceable condition is relatively high as inventory is being built up and is maintained through the procurement of new assets. Initially, when a new DLR is fielded (either as part of end-item fielding or due to an upgrade), the spare parts supply consists of all new items provided through procurement. Repair gradually becomes a greater percentage of supply as unserviceable assets of the new item start coming back from the field when they fail. We observe that, in the Other category, which consists of items with low, sporadic demand or no demand (which could reflect items that have been completely phased out), the three services look similar in terms of the percentage of assets in serviceable condition.

The Effects of Push Production

Through manual observation of demand patterns, we sought to find examples of push producing long supply conditions. We found that it is difficult to find many NIINs exhibiting clearly "wasteful" overproduction, indicating either that demand patterns have not been excessively shifting or that the exception management process has worked relatively well. This said, we provide an example to illustrate what can happen when exceptions are missed or repair continues beyond the level needed to support ongoing demands.

Figure 4.3 shows an item that has been phased out. During the phase-out process, repair continued, albeit at less than the rate of returns of the item. As a result, both unserviceable and serviceable inventory was built up. The unserviceable inventory buildup—and the disposal associated with the large drop in unserviceable inventory in mid-2007—reflects the cost of doing business when an item is phased out. However, the continuing repairs led to excess serviceable assets as well, reflecting unnecessary production and associated costs. In Appendix F, we provide three additional examples of items that exhibited excess serviceable inventory buildup during the initial phase-out period.

This is reflected at an aggregate level across the services, with push being more likely to repair into potentially excess serviceable supply. Figure 4.4 shows the percentage of DLRs, excluding dormant items and items with no serviceable inventory, with less than 0.25 serviceable inventory turns per year, which equates to four years or more of serviceable supply on hand. With the Army's more push-like system,

Figure 4.3
Example of Push Production: NIIN 011429546 Demand and Inventory History

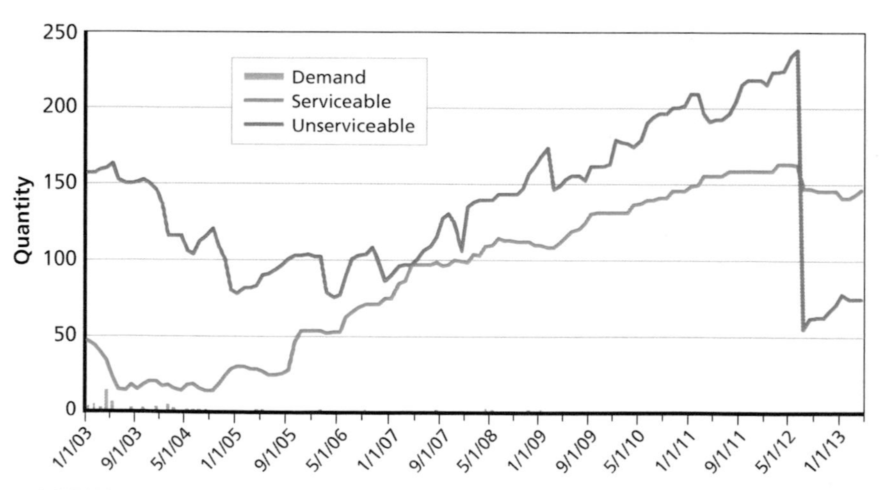

RAND *RR398-4.3*

Figure 4.4
Percentage of DLRs with Serviceable Inventory Turns Below 0.25, by Service

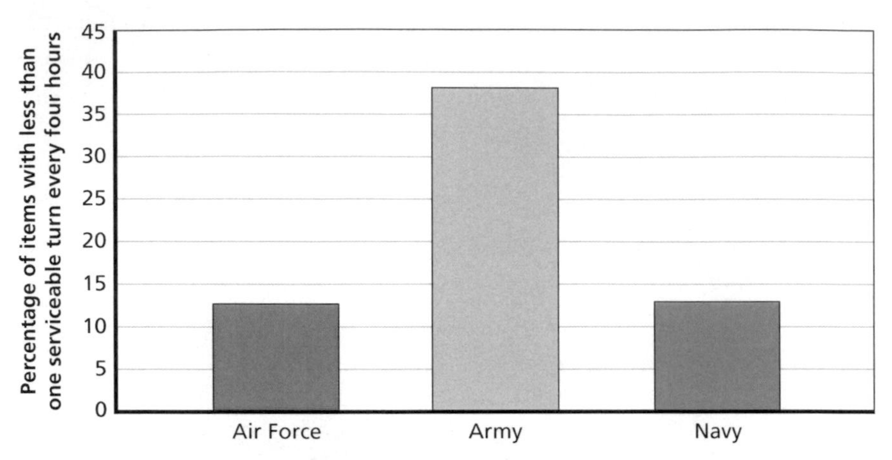

NOTE: Excludes items with no demands over ten years and no serviceable inventory over ten years.
RAND *RR398-4.4*

the percentage is higher. Not all of the NIINs for which there has been an average of four plus years of serviceable supply on hand will end up having serviceable disposals. Rather, the likelihood is higher since more serviceable stock is exposed to the risk of a further demand decrease or of demand totally disappearing due to an item being completely phased out. If demand continues instead for a relatively long period of time, then the assets will be used. Note that repairing into long supply does not increase the total assets at risk of disposal from a demand change; it only increases the risk that repairs will be unnecessarily executed.

A push system is most likely to lead to problems during a period of shifting demand. Figure 4.5 shows Army-managed item materiel availability from FY 2001 through FY 2011 based on the figures cited in annual working capital fund budgets. When demands increased in

Figure 4.5
Army-Managed Item Materiel Availability Rates, by Year

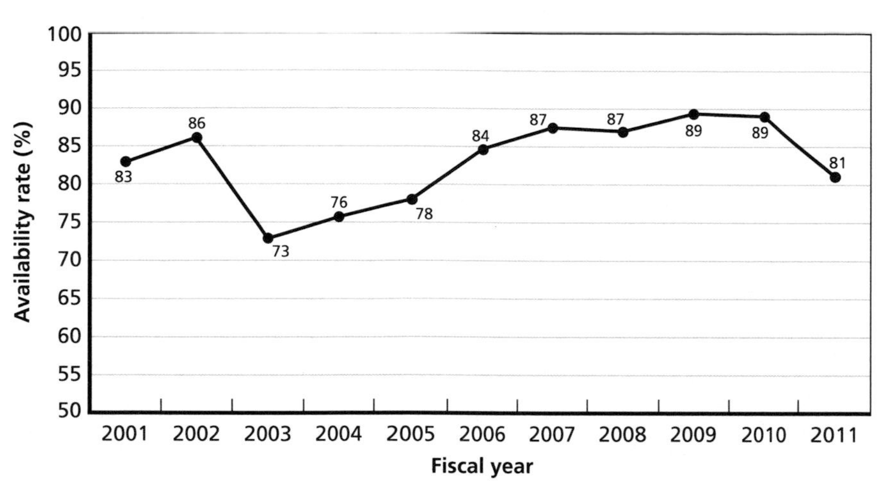

SOURCE: Army Working Capital Fund President's Budgets for Fiscal Years 2003 through 2013.
RAND *RR398-4.5*

FY 2003 with the start of Operation Iraqi Freedom (OIF), materiel availability plummeted, taking until FY 2006 to recover to close to the target at the time of 85 percent. There was a long delay in repair increases, and then the system got so far behind that it took a long period to catch up. In addition to low materiel availability, this led to situations in which repair capacity was insufficient to return serviceable inventory to sufficient levels to provide effective customer support. This led to decisions to buy more serviceable assets to improve support. This became a wartime readiness necessity, but it also increased long-term inventory, as well as potential disposals in the long term.

Push was not the only issue affecting materiel availability for the Army early in OIF. Increased demand forecasts were not proactively loaded into the planning system. Still, a pull system would have started adjusting faster, even without proactive demand forecast adjustments, because it would have called for increased production when demands started increasing instead of lagging the demand increase by a year or more. We should also note that a disruption in the retrograde pipeline from Iraq to Kuwait and then back to the United States also hampered production, as did delays in funding increases.[2] Illustrating the converse problem, Appendix F shows recent demand, repair, procurement, and inventory changes, exhibiting some lags between decreases in demand as deployed OPTEMPO declined and decreases in repairs and procurement, leading to some modest increases in serviceable inventory.

Balancing Maintenance Productivity and the Degree of Pull Production

While aligning supply and production with demand reduces inventory overall and serviceable inventory for DLRs and can improve customer support when demand shifts, it can also have implications for maintenance costs. Depot maintenance can be executed more efficiently when production is level for each line, enabling high capacity utilization. High capacity utilization can be combined with pull as well, but with typical demand variability this will necessarily result in the prioritization of items to be inducted and thus some level of back orders. And in some cases, batch production can be more efficient. Additionally, as may be indicated by the lower level of concern about parts supportability in Army depots, it can facilitate parts supportability by enabling the ordering of shortage items further in advance.

Thus, going to a completely on-demand pull system or daily planning may not always be the best total system solution. Some amount of level-loading can be important for efficiency. The degree of level-loading or push versus pull that provides the best overall result depends on the following factors:

- Repair flow time—the longer the time, the farther in advance production needs to start to meet demand and the greater the need to hold inventory to fill demands and handle variability.

[2] Unpublished research by Mark Wang, Carol Fan, Darlene Blake, Arthur Lackey, and Eric Peltz on Improving the Army's Retrograde Distribution Management Operations, 2006, and Eric Peltz, Marc Robbins, Kenneth Girardini, Rick Eden, and Jeffrey Angers, *Sustainment of Army Forces in Operation Iraqi Freedom: Major Findings and Recommendations*, Santa Monica, Calif.: RAND Corporation, MG-342, 2005.

Production needs to start in advance of demand in order to meet customer needs effectively, meriting a somewhat even flow over lead-time periods.

- The range of items made on a line or at a work station—with a larger number, a line can be better balanced in aggregate, even when the variability in demand for each item is high, enabling a greater degree of pull-like production with push-like efficiency.
- The flexibility of tooling and labor—the greater the extent to which resources can be redeployed without losses in efficiency, the more pull-like production should be.
- Space available for work in process (WIP) or AWP items and DLR sizes—if space is limited, a shop cannot afford production variability that produces WIP or to set aside in-process work waiting for parts. It needs to level-load in order to smooth workflow and ensure it is balanced across the production steps.
- Capacity utilization—a high capacity utilization activity needs to level-load; otherwise back orders will be high.
- Parts support lead time—if there are non-stocked parts that need to be ordered, greater planning lead time and thus a more push-like system will be merited.

The Air Force's ALCs demonstrate how some of these factors are considered in practice. As described, EXPRESS prioritizes inductions on a daily basis. However, during our site visits, we found a range of strategies for actually executing production, with strategies being tied to the specific characteristics of the items produced and the production facilities. At Ogden Air Force Base, there is a production facility that repairs all of the Air Force's landing gears; it produces no other items. The production of landing gears requires a large number of process steps, some with lengthy cycle times, leading to relatively long flow times of about 60 days. One of the process steps, plating, is currently at full capacity utilization. Finally, the facility has virtually no free real estate to store WIP, either in buffers between steps or from AWP conditions. As a result, the shop uses what are called M-switches in EXPRESS to provide limited level-loading of inductions, producing level-loading per day to the extent possible during a week or a month.

If, for example, a week had inductions of 5, 1, 2, 6, and 1, WIP would develop during the processing of the batches of 5 and 6, congesting the plant and further impacting production. Instead, inducting 3 each day would allow for smoother and more efficient process flow. The managers also try to avoid inducting carcasses that will be AWP during production, which drives them to intensively focus on working with DLA to improve parts supportability.

At Oklahoma City, the production of DLRs has been moved to an old General Motors auto plant that has been leased by the Air Force from Oklahoma County.[3] It is an expansive facility that provides substantial space for each production step, with significant free space remaining. Therefore, while flow times for booms and structures are long, there is space available for WIP buffers and WIP resulting from AWP. Variable daily inductions are thus executed as called for by EXPRESS, with the variability leading to expansion and contraction of the WIP. The shop thus smooths production and workload through buffer management.

Also at Oklahoma City, the engine shop employs some production smoothing. It has substantial space in the shop, so it administratively inducts engines as scheduled on a daily basis. However, given long engine-production flow times, it holds them in a preprocessing buffer, physically inducting them into maintenance in a smoother manner than called for in the demand-based induction planning.

All three shops producing items with long repair flow times apply some form of production smoothing, using techniques specific to the characteristics of the facilities. In contrast, avionics DLRs have very short repair flow times of hours rather than weeks or months. Additionally, test stands are often flexible, enabling them to be used with a variety of DLRs, and demand is relatively high in comparison to some of the items with long flow times, such as aircraft doors. These production characteristics make avionics repair conducive to daily induction planning and execution, as practiced.

---

[3] See Susan Simpson, "Tinker Employees Fill Former GM Plant," NewsOK.com, August 22, 2010.

Moving Toward Pull Production for the Army

Prior analysis for the Army documented in unpublished research suggests that much of the gains from shifting to a pull system could be garnered through three-month planning horizons, or even six-month horizons, similar to current Navy practices.[4] These periods are long enough to accommodate most parts planning windows and allow for some production smoothing while still reacting in one-fourth to one-half the time of annual workloading, and changes in workload plans would not rely on exception management. A comparison between the Air Force and Navy supports this conclusion, with relatively similar levels of serviceable asset percentages and low serviceable excess for the Navy.

Thus, as discussed in the overall recommendations in Chapter Eight, we recommend that the Army adopt modified pull production with the aim of increasing workloading frequency and reducing the time horizons of firm orders passed to the depots. It could start by shifting all production planning to six-month intervals. After a period of learning and working through resulting issues, it could then further decrease the intervals to four-month, three-month, or even shorter intervals on a customized basis tailored to the characteristics of NIIN-specific demand and repair profiles and production facility characteristics.

---

[4] Unpublished research by Mark Wang, Jason Eng, Rachel Rue, and Jeffrey Tew on Adapting Army Secondary Item Planning to Pull Production, November 2009.

CHAPTER FIVE

Reducing Contract Lead Times

Efforts to reduce contract lead times would improve DLR supply chain management in three different ways.

First, the longer the procurement lead time for new DLRs, the greater the potential inventory excess and customer support impacts of unanticipated changes in demand. This is because the longer the lead time, the greater the risk of a demand shift from the plan occurring within the lead-time window. Such a shift can produce excess inventory

- if the forecasted demand for a new DLR is greater than the realized demand, which can occur if the fleet size, OPTEMPO, or reliability is different than planned/forecasted
- if a planned increase in demand is cancelled or otherwise does not materialize.

Conversely, longer lead times contribute to customer support problems if demand increases due to unanticipated conditions or events.

Second, a confluence of three factors can lead to long repair-contract lead times, in turn leading to excessive assets or poor customer support. These three factors are

- insufficient assets in the system to support demand through repair in the face of a demand shift or even possibly with stable demand

- a standard plan to rely on contract repair for some or all capacity[1]
- not having a contract in place. This could stem from the following general situations:
  - when repair has not been occurring because there was excess supply of serviceable assets, a repair contract expired (or was not in place), and then it was not renewed in anticipation of some future demand. This would come into play when demand for repair restarts from zero or when there is a heavy increase in demand.
  - when production has been occurring, but the renewal of the repair contract runs into unexpected issues.

When there is a delay in establishing a repair contract that delays inductions of assets, this leads to customer support problems. If this delay becomes severe, the item manager may need to order new assets to serve customers, producing excess assets in the system in the long run.

Third, lead times for piece parts used in repairs can affect AWP time and production efficiency. Changes in DLR demand and repair forecasts result in changes in indentured part needs. This could result in the need for increased buys. Depending on the length of the advance notice of the change and the part lead time, this could lead to stockouts in support of maintenance.

---

[1] In our interviews, we found that the services use contract repair as a second source of repair on some items to absorb variability in demand, when needed for sufficient total capacity, or as the primary source of repair when capabilities or data rights so dictate or when found to be less costly. There are also times when the item managers use an existing repair contract setup as a second source of repair when organic parts support falls short. It is reported that, in these cases, the repair contractor is often able to get the needed parts and conduct the needed repairs. This has not been measured, though.

Example Effects of Long Lead Times/Contract Renewal Problems

We illustrate these situations with a few examples. The first is an Air Force–managed part for the F-15 landing gear. The F-15 system program office (SPO) had decided to make landing gear replacement a mandatory action in the PDM for F-15s instead of using on-condition replacement. To accommodate this change, which would increase demand for landing gears, additional parts to support landing-gear production were ordered. The example item in Figure 5.1 had approximately a four-year lead time. Between the time it was ordered and the planned start of the PDM change, which was tied to when sufficient parts would be available for production, a program budget shortfall led to the cancellation of the PDM change. But it was too late to economically cancel the contract, leading to delivery of what became excess assets. This is seen in Figure 5.1 by the climb in serviceable assets in 2011 and 2012. If the lead time for the purchase of new assets had been shorter, the change may have occurred before the order had to be placed to meet the anticipated need.

It can also be seen in Figure 5.1 that there were periods of zero or virtually no serviceable inventory of this item to support customers, while large numbers of unserviceable assets were available for repair. This stemmed from parts shortages, which led to a decision to temporarily stop inducting landing gear pistons to avoid building up WIP of partially repaired pistons.

In the next example, shown in Figure 5.2, production had been halted due to long serviceable supply of the asset, which shows a climb in serviceable inventory in 2007. As this item's serviceable inventory was drawn down, the need for repair returned. However, the repair contract expired in 2009. Delays in initiating a new contract led to customer support problems, as reflected by the increased ICP processing time, or back-order time, in the fulfillment of customer orders (see Figure 5.3). The fleet size was also increasing, so some additional new assets were needed to meet the higher level of demand. But to help compensate for the support problem, additional items were ordered, increasing total system assets and the potential for long-run excess.

Figure 5.1
Long Lead Time Excess Example: NIIN 010803407 Demand and Inventory History

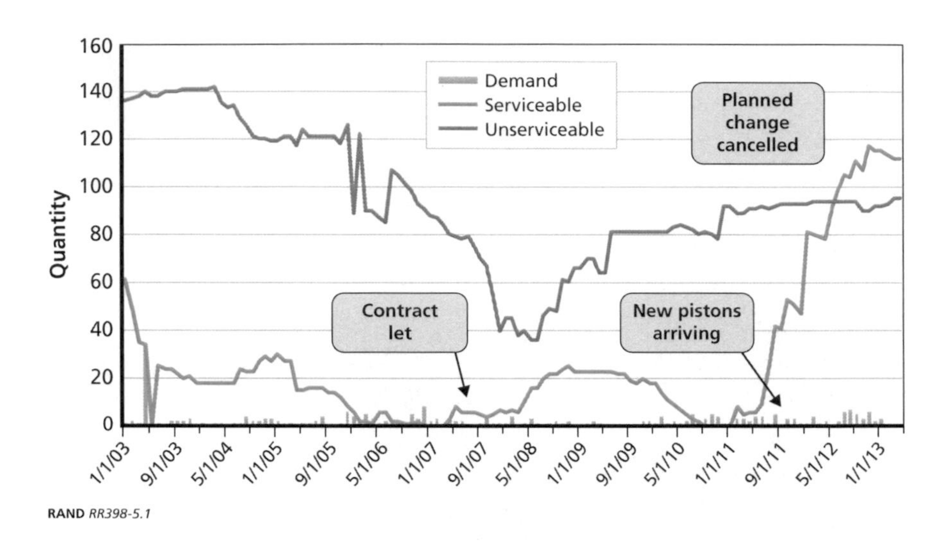

RAND *RR398-5.1*

Figure 5.2
Repair Contract Delay Example: NIIN 012844013 Demand and Inventory History

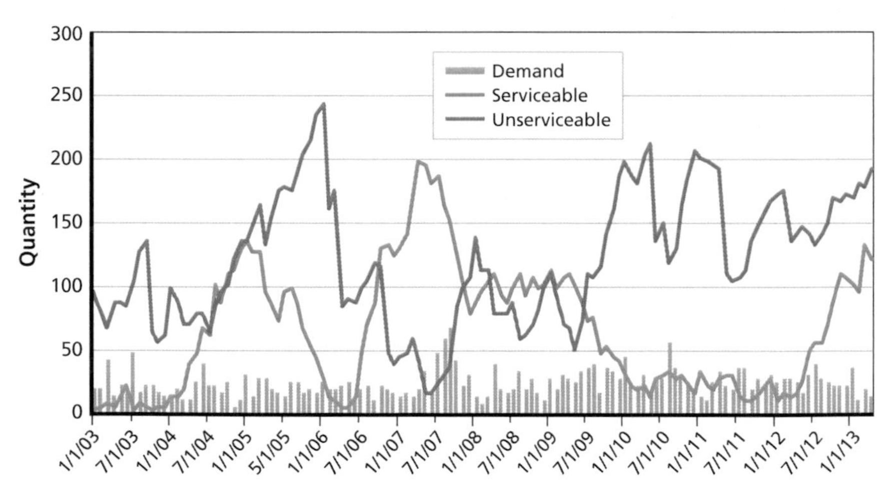

RAND *RR398-5.2*

Figure 5.3
NIIN 012844013 ICP Time and New Buys History

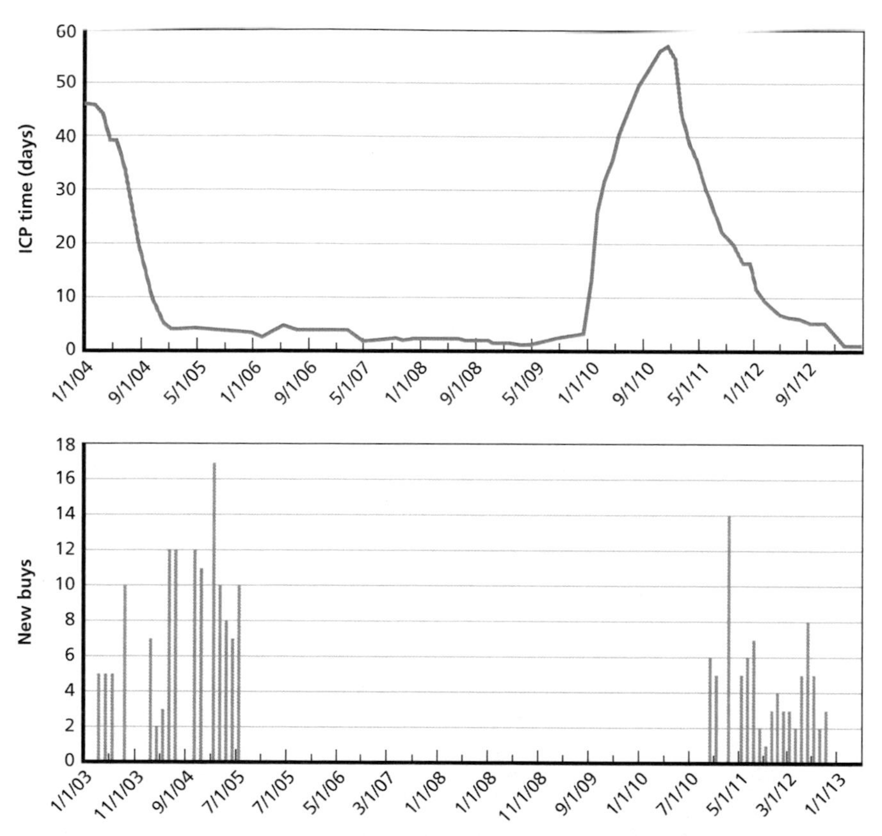

NOTE: ICP time is the time from the receipt of the requisition by the ICP until the materiel release order.

RAND *RR398-5.3*

CHAPTER SIX

Improving Repair Planning: Anticipating Knowable Changes in Demand and Condemnations

Changes in demand for repair, either due to demand shifts or shifts in condemnation rates, can occur for anticipatable reasons that are not automatically captured in DLR management information systems and that are not always accounted for by item managers. There is a small subset of DLRs that can only be repaired a fixed number of times before the asset can no longer be renewed, leading to automatic condemnation and disposal. If this occurs during a somewhat narrow window across the fleet, condemnations can abruptly spike and prevent effective support through repair only. There will not be enough carcasses to repair. New assets will have to be purchased to replace the condemnations. If the item manager does not anticipate this and manually plan for this by changing, for instance, the condemnation rate in the planning system starting at the right point in the future, customers can be left with poor support during the lead time for buying new assets.

Figure 6.1 shows an example of this situation. The item in Figure 6.1 can only be repaired three times before disposal, with low condemnations before the repair limit is hit. The repair limits were hit in a narrow window across the fleet, with three spikes of condemnations in 2007, as shown in the top graph of Figure 6.1. These were not fully anticipated, leading to new buys that fell short of fully replacing the assets until 2011. The middle graph shows the average monthly back-order time resulting from the shortage of assets. The bottom graph shows receipts of new buys per month, with new deliveries not fully alleviating back orders until October 2012, after five years of subpar support.

Figure 6.1
Example Lifetime Repair Limit Item: NIIN 011117804 Condemnations Leading to Disposals, ICP Time, and New Buys History

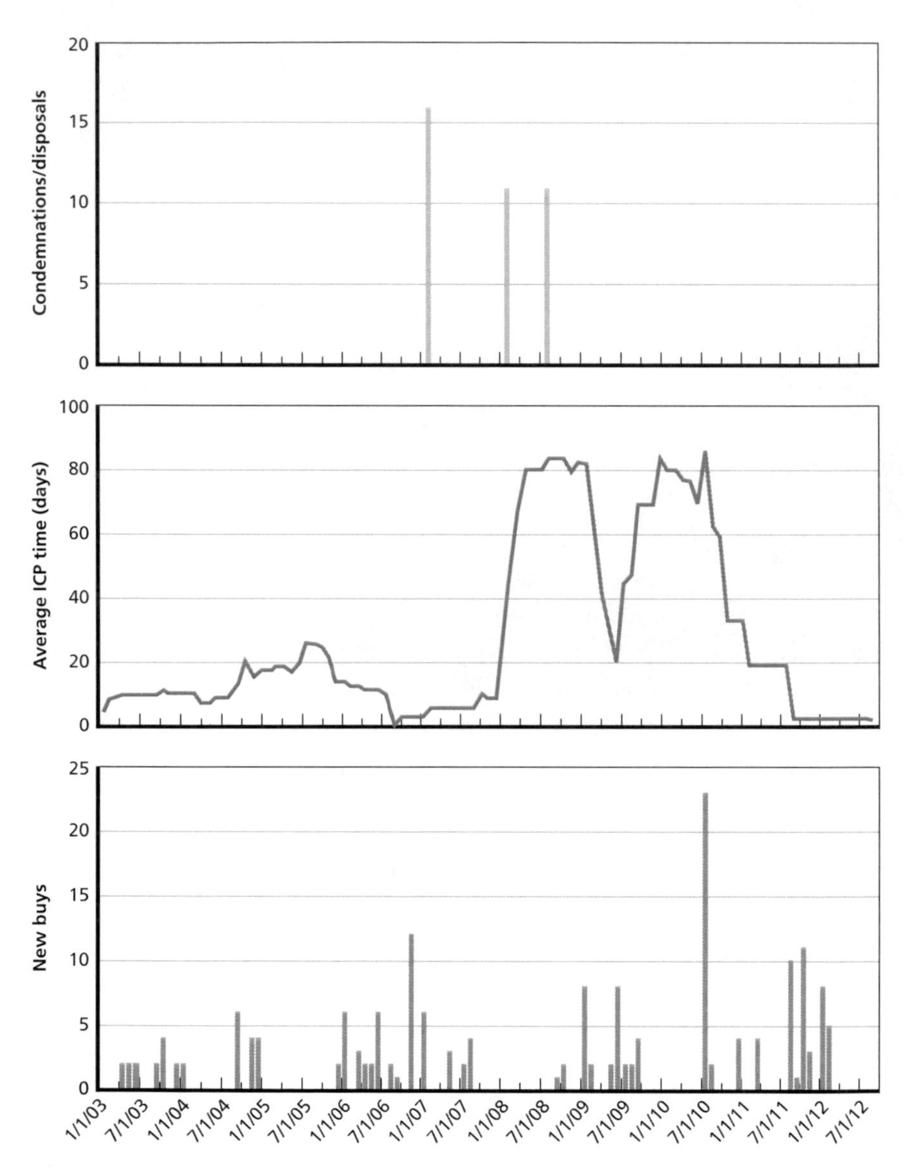

RAND *RR398-6.1*

A similar situation can occur early in a DLR's life cycle. If a DLR is phased in quickly and it is a wear-out item, demands will be low until the first "wear-out cycle" is reached. Demands then suddenly increase, potentially leading to resource shortage problems in support of production. In this case, the shortage is not of new DLRs, but of the resources to conduct repair, such as indentured parts. This situation or any other that temporarily eliminates the need for repair can also produce a procurement delay problem for an item with a relatively high condemnation rate. For example, the Air Force maintenance management system only tracks condemnation history for the prior two years. If no repairs occur during a two-year period, it will revert to no condemnations. When repair then starts, this planning factor could be off.

More generally, when repair has been temporarily halted or is below the ongoing level of demand, the system can be left short of assets to reinitiate repair altogether or at the higher level of demand. The maintenance planning systems use repair lead time, comprising shop flow time, to determine when to start inducting assets in order to have them in serviceable condition when needed. The systems also use procurement lead time to determine when new buys need to be placed in the face of demand increases or condemnations. But repair requires labor and parts. If these are not both on hand when repair induction is called for, repair cannot start or be completed. The planning system ideally would anticipate the need for these resources a lead time before the needed induction date. This could be done in two ways. The first is to record the lead time for all supporting resources in the planning system, having the resource with the longest lead time producing the action-initiation lead time. The second is for the item manager to manually determine this and plan accordingly, for instance, by developing DDEs for DLA to purchase needed parts in anticipation of production.

CHAPTER SEVEN

Potential DLR Management Changes for Exploration from the Literature

A review of practices in the aviation industry and other industries that use reparables or rotable spares in the private sector may provide useful information for policymakers seeking to enhance DLR item management by the services. In this chapter, we consider how private enterprise has dealt with critical issues such as parts supportability and obsolescence, as well as potential opportunities for process and efficiency improvements. However, we found that there is limited detail available on actual item management practices, with most available information pertaining to overall high-level supply chain issues or inventory planning methods, which were not the focus of this study. Nevertheless, we found that strong parallels exist between the management of DLRs by the services and by industry. The private sector faces similar challenges, including the "DLR" disposal cost of doing business, which is only exacerbated for DoD by longer product life cycles.

One of the key issues that the services contend with is the volatility and irregularity of demand for DLRs, a problem that the private sector often confronts as well. In the airline sector, typically more than 50 percent of rotable inventory does not move over a two-year period.[1] According to an inventory control specialist in the airline industry, this apparently low turn rate "isn't too bad" and is essentially a cost of doing business. Olivetti S.p.A., an Italian PC manufacturer, also faces a similar demand pattern, with just 20 percent of reparable compo-

---

[1] Kieran Daly, "Cash on the Shelf," *Airline Business*, Vol. 25, No. 11, November 2009, pp. 62–63.

nents representing 80 percent of demand.[2] Technology firms such as IBM often confront short product life cycles, leading to a sudden drop in demand for spares and the subsequent buildup of spare inventory.[3] These unbalanced and volatile demand patterns lead firms to maintain larger safety stocks, perhaps not reflecting well on the balance sheet.[4]

Parts support and obsolescence are critical issues for private firms, just as they are for the services. As discussed in a recent *Aviation Week* article, executives at US Airways have struggled with maintaining avionics systems on the airline's legacy Boeing 737-300s and 737-400s.[5] The turnover rate for avionics systems often far outpaces the turnover rate of the aircraft on which they are installed, complicating maintenance and repair for US Airways and other carriers. As new avionics systems are introduced, niche specialists tend to move toward supporting new systems and away from legacy systems, according to the article. This flow of labor hampers the ability of the airlines to complete timely repairs.

To confront these issues, firms have pursued a variety of strategies. Negotiating longer-term maintenance and repair contracts at the beginning of a product life cycle can help alleviate parts supportability issues later in the product life cycle.[6] Original equipment manufacturers (OEMs) have also started to offer the airlines the ability to enroll in total life cycle management programs, taking greater responsibility for maintenance and repair, as well as stocking spares.[7] With the rollout of the Boeing 787 Dreamliner, Boeing began its GoldCare program,

---

[2] J. R. Ashayeri et al., "Inventory Management of Repairable Service Parts for Personal Computers: A Case Study," *International Journal of Operations & Production Management*, Vol. 16, No. 12, 1996, pp. 74–97.

[3] M. J. Fleischmann et al., "Integrating Closed-Loop Supply Chains and Spare-Parts Management at IBM," *Interfaces*, Vol. 33, No. 6., 2003, pp. 44–56.

[4] Daly, 2009.

[5] Paul Seidenman and David Spanovich, "Reconfiguring Avionics Support," *Aviation Week Overhaul and Maintenance*, Vol. 18, No. 5, May 2012, pp. 18–22.

[6] Daly, 2009.

[7] Robert W. Moorman, "OEMs Tout Savings Via Rotable Programs," *Aviation Week*, Vol. 17, No. 1, January 2011, p. 52.

providing an opportunity for airlines to receive total life cycle management services.[8] Airbus has a similar program called Flight Hour Services, which British Airways recently enrolled in to maintain its fleet of A380 aircraft.[9] Like OEMs, third-party repair firms offer maintenance and repair service contracts, assuming the risk of potential stockouts.[10] These practices are consistent with recommendations made in other RAND research for longer contracts and purchasing support in conjunction with end items.[11] There is also some similarity with some DoD performance-based logistics contracts such as those for the F-22, C-17, and F-117, with the contractor being responsible for managing the supply chain and overall sustainment.[12] However, some portion of depot-level repair work has typically been kept within government depots in these contracts due to statutory core logistics requirements to maintain critical capabilities in the government base and the requirement for at least 50 percent of depot work be done in house.[13]

---

8 M. Mecham, "Golden TUI," *Aviation Week and Space Technology*, Vol. 172, No. 15, 2010, p. 45.

9 Airbus, "British Airways Selects Airbus Flight Hour Services for Its A380 Fleet," October 6, 2011.

10 Harold Krikke and Erwin van der Laan, "Last Time Buy and Control Policies with Phase-Out Returns: A Case Study in Plant Control Systems," *International Journal of Production Research*, Vol. 49, 2011, pp. 5183–5206.

11 Mary E. Chenoweth, Jeremy Arkes, and Nancy Y. Moore, *Best Practices in Developing Proactive Supply Strategies for Air Force Low-Demand Service Parts*, Santa Monica, Calif.: RAND Corporation, MG-858-AF, 2010.

12 Cynthia R. Cook, Michael Boito, John C. Graser, Edward G. Keating, Michael J. Neumann, and Ian P. Cook, *A Methodology for Comparing Costs and Benefits of Management Alternatives for F-22 Sustainment*, Santa Monica, Calif.: RAND Corporation, TR-763-AF, 2011; "The Global C-17 Sustainment Partnership," *Defense Industry Daily*, January 7, 2013; Michael Boito, Cynthia R. Cook, and John C. Graser, *Contractor Logistics Support in the U.S. Air Force*, Santa Monica, Calif.: RAND Corporation, MG-779-AF, 2009.

13 U.S. law requires that at least 50 percent of depot maintenance be executed by government employees (United States Code, Title 10, Section 2466, Limitations on the Performance of Depot-Level Maintenance of Materiel). U.S. law also requires that DoD maintain government-owned and -operated technical capabilities necessary to ensure the ability to meet wartime needs (United States Code, Title 10, Section 2464, Core Depot-Level Maintenance and Repair Capabilities).

To manage short product life cycles and item phase-outs, IBM has been a pioneer in the field of reverse logistics. Fleischmann et al. assess how IBM exploits returned, used, and irreparable machines as a source of spare parts. The authors discuss the trade-offs involved with dismantling end items and reusing spare parts. Some parts may be dismantled that are eventually not needed, resulting in added cost. Conversely, some parts that would have been very cheap to obtain through dismantling can be missed, leading to potentially expensive new buys or difficulties in obtaining new parts. According to the authors, good communication and organizational visibility are essential to ensuring the viability of a reverse logistics program. To increase the opportunities for reclamation of spare parts, IBM has created a program through which customers can recycle old machines.[14]

Finally, software that provides total asset visibility, as well as modeling and forecasting capabilities, has been particularly beneficial for several firms in identifying potential problems. Tedone describes the mathematical optimization modeling used for American Airlines's Rotables Allocation and Planning System (RAPS).[15] The software allows users to forecast demand and determine the least cost allocation of spare parts across airport locations. When rolled out in the late 1980s, RAPS was estimated to provide a one-time savings of $7 million and an annual savings of approximately $1 million for American Airlines, according to Tedone. In addition, Tedone notes that RAPS has increased the productivity of item managers, allowing them to manage a much greater number of items each day. Such modeling software also proved successful for Cummins Engine Company, which manufactures large and expensive engines.[16] To provide repair service to customers, Cummins relied on a network of local dealers who were very small and had difficulty managing inventory for over 15,000 parts. By using modeling software, Cummins was able to increase customer service

---

[14] Fleischmann et al., 2003.

[15] Mark J. Tedone, "Repairable Part Management," *Interfaces*, July–August 2009, Vol. 19, No. 4, pp. 61–68.

[16] H. L. Richardson, "Service Parts: Reaching the Right Level," *Transportation & Distribution*, Vol. 39, No. 9, 1998, p. 39.

levels and lower inventory by 20–30 percent. This is consistent with DoD information system practices, such as the Air Force's EXPRESS system, the Navy's Re-Engineered Maritime Allowance Development (ReMAD), and the adoption of ERPs.

While providing limited process improvement suggestions beyond those identified in the analysis of DoD DLR management described in this report, this review does suggest a couple of broader concepts to explore in terms of overall DLR management. One such concept would be dedicated harvesting operations for when items are phased out. The value of this would depend on whether DLRs share common parts, which would need to be examined. If they do, such parts could be harvested from phase-out items for use on other DLRs.

Another insight is that additional pooling or consolidation within and across services around common technology operations might lead to maintenance infrastructure efficiencies. Existing examples include the consolidated shops for Air Force landing gears and engines mentioned earlier in this report. Such consolidation is relatively common within the services but is less pervasive across the services. Again, determining whether efficiencies would accrue would require additional research beyond the bounds of this project and report.

CHAPTER EIGHT

Overall Conclusions, Recommendations, and Needs for Further Research

DLR supply chain management appears to be done relatively effectively across the services. We did not identify any single process improvement opportunity for dramatically reducing inventory requirements. However, improved processes for providing needed parts to depot maintenance would improve customer support and could reduce total system costs through improved maintenance productivity.

It is difficult to find NIINs with excess inventory due to avoidable situations or poor supply chain management decisions or processes. Instead, the buildup of excess and resulting disposals stems primarily from DLR phase-outs due to DLR upgrades and fleet size reductions, combined with very low condemnation rates. DoD should better explain the item–phase-out impact on DLR inventory to improve understanding among external stakeholders and should isolate excess DLR inventory and disposals within overall inventory and disposal metrics to make them directly visible. When items are replaced or end-item fleets are phased out, the plans should include the long-term disposal plans for DLRs. We observe that, in most cases in which excess builds up due to phase-outs and leads to disposals, there is a multiyear delay. It appears likely that these delays could be shortened, potentially reducing storage costs and warehouse infrastructure requirements. Disposals could begin before an item is completely phased out as the inventory becomes greater than that needed to effectively maintain readiness and the readiness risk of disposing of some assets becomes quite low. Disposing of DLRs more quickly after phase-out and/or disposing of them in phases or steps could also improve perceptions of DoD inventory management by increasing inventory turns, reducing excess, and reducing overall inventory. Similarly, when looking at inventory turns, phase-out items should be separated out to provide a better understanding of turns for steady-state items. Focusing on steady-state items will provide a better understanding of how closed-loop DLR process improvement would affect inventory requirements for new DLRs, as well as provide a better understanding of whether supply chain management improvement efforts are producing progress.

Beyond this, the one major practice gap in DLR supply chain management that hinders DoD's agility to align supply with demand to provide effective customer support and avoid building up excess is the Army's use of a push-like production system. The Army should take steps to move toward a more pull-like paradigm, starting with adopting workload planning horizons that are no longer than six months and then working to shorten these horizons shop-by-shop, as merited through process learning and continuous improvement efforts, based on shop-specific demand and facility characteristics. More broadly, the services should seek to find the workload planning horizons that minimize total costs when considering supply and maintenance, seeking to shorten the horizon without impacting maintenance productivity. Shorter horizons enable faster response to changes in demand and reveal process inefficiencies.

A broader issue that impacts DLR supply chain management customer support effectiveness and efficiency is parts supportability. The biggest direct impact of this issue is degradation in customer support when severe parts shortages lead to the complete depletion of serviceable wholesale inventory of a DLR. The costs accrue on two fronts. Part delays and variability in wait times increase the closed-loop repair cycle time, increasing steady-state inventory requirements. These delays can also influence service selection of pull-oriented planning horizons—i.e., where the service selects to be on the push-pull spectrum—and thus the responsiveness of the DLR supply chain, further affecting inventory and customer support. The second front includes the impacts on maintenance productivity and on the costs of workarounds executed by the maintenance depots to compensate for parts shortages.

Such costs could be substantial but are not directly measureable given current data collection.

There are several paths to improved parts supportability and overall supply chain integration. First, the services could quantitatively estimate the costs of parts shortages and DLA could incorporate these costs in safety stock planning to jointly trade off these costs to determine the balance and associated level of inventory that produces the lowest total cost for DoD as a whole. Second, continued improvements could be made in collaboration and coordination. Effective service processes for sharing planning data with DLA in an actionable form are key. This includes both automated data sharing, as feasible, and the manual provision of useable information. These processes need to be designed to ensure all planning information that could improve parts forecasts is shared. Of note, the Navy and Army are in the process of linking their ERPs that develop DLR production plans with DLA's ERP for automated sharing of DLR production plans and associated part needs. In conjunction, having the right metrics and using them for performance-based agreements between the services and DLA, with the office of the ASD(L&MR) establishing a standard base template, would provide a communications tool and better ensure that interorganizational priorities are aligned, driving the effective execution of both collaborative planning and intraorganizational planning processes. Finally, when the planning processes that warn of repair-plan changes break down or unexpected changes in demand occur, reduced lead times for procuring piece parts can reduce awaiting parts time.

A potential avenue to further explore is how DLR repair contractors manage parts support and whether this offers lessons for improved collaboration between the services and DLA or improved supply and inventory planning. Another possibility is that this could also reinforce the need to consider the costs of stockouts on maintenance, particularly given the profit motivation of private sector firms.

Lead times also impact DLR customer support effectiveness and efficiency in two other ways. One is the lead time for repair contracts, which becomes important when a long repair contract lead time leads to a lapse in a contract or when a new repair contract is needed due to a change in demand. A new contract might be needed if demand suddenly surges or new faults or conditions appear due to new usage conditions or the development of new fault types late in a DLR's life cycle. The second way is when a demand increase requires an increase in total inventory of a DLR. In this case, the lead time for procuring new DLRs could affect customer support depending on how it compares to the warning lead time. Additionally, if there is a planned increase in demand that triggers increased buys of both a DLR and its indentured parts, but the plan does not materialize, this will lead to excess inventory. The longer the procurement lead times, the greater risk there is of this occurring.

In addition to parts planning improvements, there is some potential for repair planning improvements. The service planning systems determine when repairs need to be initiated based on the repair flow time. This assumes that repairs can be initiated upon induction and that there will not be delays caused by supporting resources. However, particularly when an item has not been in production for some period, not all of the requisite resources are always immediately available. These resources include parts and repair contracts. If lead time is required to get these resources in advance of starting a repair, the systems need to flag this as well, kicking off the purchasing process for parts or a repair contract in advance. Ideally, this would be automated in the planning systems. But until this is possible, item managers should check these conditions periodically for items not currently in production, looking ahead to when production is projected to start, checking the supporting parts inventory for the existence of a repair contract, as appropriate to the specific DLR. If parts are insufficient or a needed repair contract is not in place, the item manager should ensure actions to procure these resources are initiated a lead time in advance of future repair.

Similarly, item managers need to pay special attention to new items and lifetime-repair–limited items, tracking installation periods and the number of lifetime replacements, respectively. From this, they should track when they would expect demands and condemnations, respectively, to increase. In the former case, a repair and potentially a procurement increase will need to occur. In the latter case, a procurement will be needed to compensate for the temporary increase in the condemnation rate.

Finally, the case studies described in Chapter Two were essential for identifying the causes of DLR inventory excess and customer support shortfalls. However, the number of case studies that could be coordinated and accomplished as part of the project was limited to a couple dozen. While these were sufficient to identify a range of causes and give a reasonably strong indication as to which are the most important, a larger sample would potentially uncover additional process issues and could better identify the critical issues for each service and specific types of DLRs. Also, as our recommendations or other process improvements are implemented, new case studies could help provide feedback on process change effectiveness. So we recommend that the NIIN case study approach be employed as part of future process improvement efforts.

Additionally, certain situations could be automatically flagged when they occur for this type of case study analysis. For example, if a critical customer support shortfall forces a procurement action when significant unserviceable inventory of a DLR is on hand, it could be recorded and could be a trigger to identify the constraint on production that forced the purchase of additional inventory. Identifying and recording the root cause of a problem when it occurs takes much less work than going back later to conduct historical case studies, like those executed for this report. Also, tracking the occurrence of such events over time could be useful for monitoring progress. Similarly, the accumulation of data on production constraints would lead to the identification of key process shortfalls hampering system efficiency.

Beyond the process disruption events and overall production planning approaches that are the focus of this report, the services have long sought to improve DLR processes, focusing to a great degree on repair process flow time and on retrograde time in the Navy and Air Force.[1] Reducing standard process time reduces the amount of inventory the

---

[1] For example, a recent three-pronged approach by the Navy has led to improved efficiency and accountability in the retrogade process (Electronic Retrograde Management System), shorter retrograde times (Advanced Traceability and Control), and better classification and packaging for improved efficiency and quality (Technical Assistance for Repairables Processing). Naval Supply Systems Command, "Navy Retrograde to Army G-4," briefing, January 27, 2011.

closed-loop system needs to meet customer needs, which reduces initial buy requirements. In turn, when an item is phased out, this will be reflected in reduced disposals of economically useful assets. Such process improvements should be ongoing in perpetuity in the spirit of continuous improvement. In addition to improving maintenance efficiency, this will reduce DLR purchase requirements each time a new DLR is phased in or a new system is fielded.

APPENDIX A

Air Force DLR Management

Organization

AFMC, headquartered at Wright-Patterson Air Force Base in Dayton, Ohio, oversees five centers, as shown in Figure A.1.[1] AFMC headquarters provides overall planning and oversight for Air Force weapon system acquisition and sustainment. AFMC conducts research, development, test and evaluation, and provides the acquisition management services and logistics support necessary to keep Air Force weapon systems ready for war.[2] Within AFMC, the Air Force Life Cycle Management Center (AFLCMC) supports the Air Force acquisition executive with weapon system product development and support system design functions. The Air Force Sustainment Center (AFSC) provides depot-level weapon system maintenance and Air Force supply chain management functions. The Air Force Test Center (AFTC) provides weapon system test and evaluation functions. The Air Force Research Laboratory (AFRL) provides research and development functions. The Air

[1] Figures A.1 and A.2 are derived from Robert S. Tripp, Kristin F. Lynch, Daniel M. Romano, William L. Shelton, John A. Ausink, Chelsea Kaihoi Duran, Robert G. DeFeo, David W. George, Raymond E. Conley, Bernard Fox, and Jerry M. Sollinger, *Air Force Materiel Command Reorganization Analysis: Final Report*, Santa Monica, Calif.: RAND Corporation, MG-1219-AF, 2012.

[2] See U.S. Air Force, "Air Force Materiel Command," January 30, 2013.

Figure A.1
Air Force Materiel Command Centers

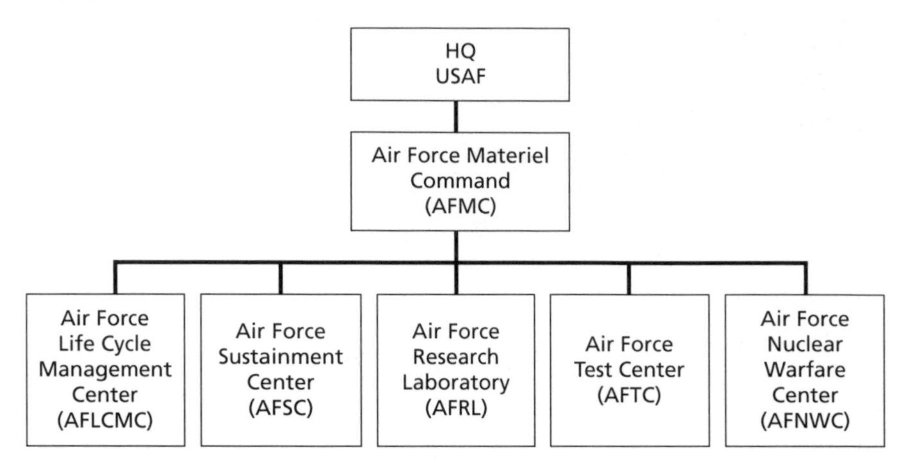

RAND *RR398-A.1*

Force Nuclear Warfare Center (AFNWC) focuses on nuclear weapon functions.[3]

The AFSC's mission is to sustain weapon system readiness through depot maintenance, supply chain management, and installation support.[4] The AFSC is further divided into three Air Logistics Complexes (ALCs) and two supply chain wings, as shown in Figure A.2. While the ALCs have similar organizational structures and missions, each performs depot-level maintenance on different weapon systems. The 448th Supply Chain Management Wing (SCMW) is responsible for supply planning for depot-level repairs (e.g., accurate forecasting, efficient sourcing, and improving system and equipment availability).[5] The 448th SCMW has supply chain management groups at each ALC responsible for supply chain requirement planning and execution for that complex and for different commodity groups. The 635th Supply

---

3 See U.S. Air Force Materiel Command Public Affairs, "AFMC Restructures to Cut Overhead, Make Command More Efficient," November 18, 2011.

4 See U.S. Air Force Sustainment Center, "Air Force Sustainment Center," July 20, 2012.

5 See Tom Girz, "448th Supply Chain Management Wing: Long Term Strategic Plan Update," briefing, January 25, 2011.

Figure A.2
Air Force Sustainment Center Organizational Structure

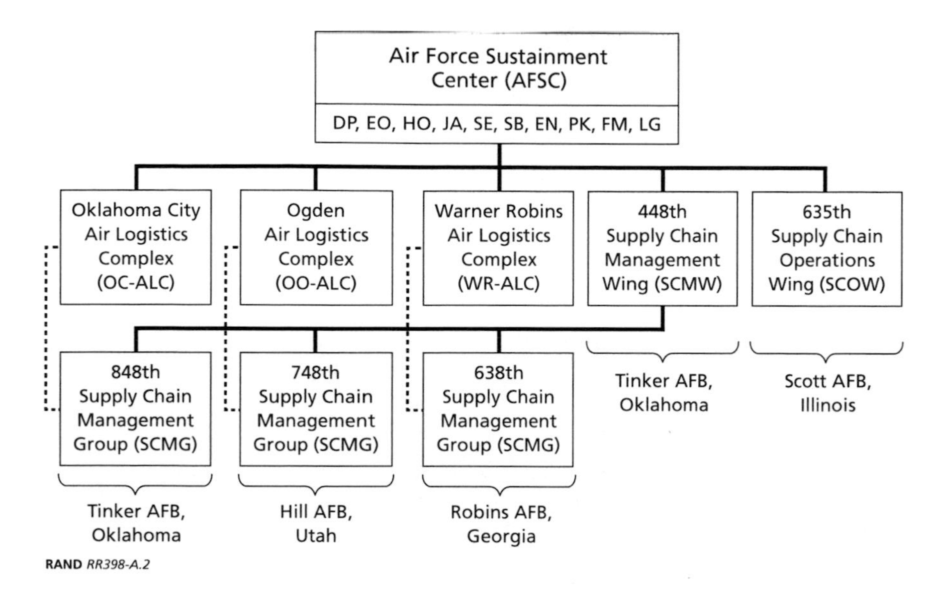

RAND *RR398-A.2*

Chain Operations Wing (SCOW) takes an outward focus, supporting Air Force weapon systems worldwide through resource allocation prioritization and linking readiness needs to providers.

Each ALC is organized into groups, as shown in Figure A.3. As seen in Figure A.3, some maintenance groups are only found at specific ALCs where there is a concentration of expertise and repair capability, e.g., all aircraft engines are maintained by the 76th Propulsion Maintenance Group at the Oklahoma City Air Logistics Complex (OC-ALC). These differences are related to the ALCs' areas of emphasis, as summarized in Table A.1, which provides an overview of primary capabilities and operations rather than an exhaustive list.

Maintenance Groups

Each depot's aircraft and commodity maintenance groups are responsible for the hands-on maintenance of aircraft or other "end items," [6] as

---

6 Aircraft engines are traditionally viewed as end items even though they are components of an aircraft.

Figure A.3
Air Logistics Complex Groups

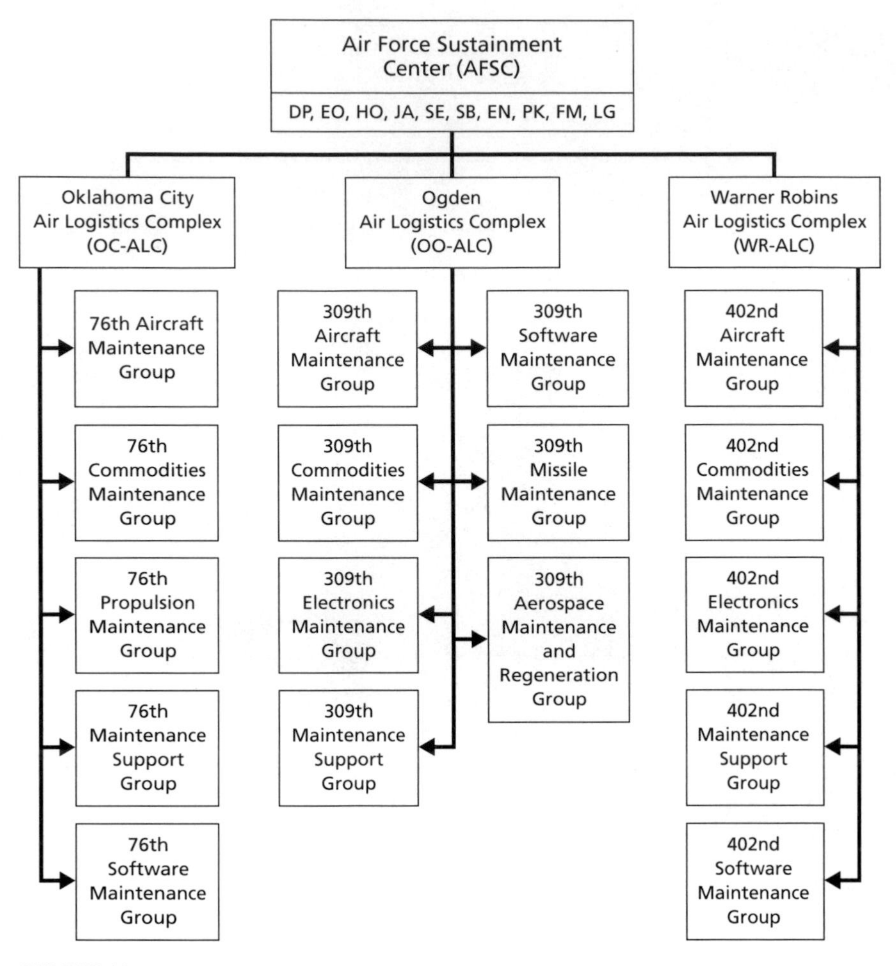

RAND *RR398-A.3*

well as for the repair[7] of items that are components of end items. Broken parts are sent to depot maintainers, who are then asked to repair them. There can be considerable indenture. For example, a landing gear is a

---

[7] *Maintenance* and *repair* are synonyms in this context.

Table A.1
A High-Level Summary of Air Logistics Complex Emphases

Category	Ogden	Oklahoma City	Warner Robins
Traditional Foci	Fighter aircraft, landing gear	Aircraft engines	Cargo aircraft
Depot-Level Maintenance/ Modification Aircraft	A-10, F-16, F-22A, C-130	B-1, B-2, B-52, KC-135, E-3 AWACS	C-5, C-17, C-130, F-15, Air Force helicopters, Special Operations Forces aircraft
Major Systems	Landing gear, wheels and brakes, low observables, ICBM support	F100, F101, F108, F110, F117, F118, F119, TF33 engines	Avionics, electronic warfare

SOURCES: Allan Day, "309th Maintenance Wing Overview," Ogden ALC 309th Maintenance Wing, August 16, 2011; Hill Air Force Base, "U.S. Air Force Fact Sheet: AFLC," undated; Ogden Air Logistics Center Public Affairs Office, "Ogden Air Logistics Center," undated; Robins Air Force Base, "Warner Robins Air Logistics Complex," undated; Tinker Public Affairs, "U.S. Air Force Fact Sheet: Oklahoma City Air Logistics Complex," August 28, 2012.

component of an aircraft, but a landing gear itself has hundreds of subcomponents, comprising both indentured reparables and consumables.

Maintenance can be either unscheduled or scheduled, including PDM, in which an aircraft receives an extensive refurbishment that may involve large-scale disassembly; refurbishment, replacement, and upgrades of parts; and ultimate reassembly before return to service. Aircraft PDM visits can be lengthy, covering scores of days[8] and tens of thousands of labor hours. Aircraft modification/upgrade programs are another form of scheduled depot-level maintenance.

While the depots have traditional areas of specialization, as noted in Table A.1, all three depots might be involved in a given aircraft's PDM visit or upgrade program. A C-5 cargo aircraft that goes to the Warner Robins Air Logistics Complex (WR-ALC) for PDM would have its engines removed and sent to OC-ALC and its landing gear

---

[8] For instance, Warner Robins F-15 PDM visits averaged between 100 and 130 days between FY 2000 and FY 2006 (Keating, Resnick, Loredo, and Hillestad, *Insights on Aircraft Programmed Depot Maintenance: An Analysis of F-15 PDM*, Santa Monica, Calif.: RAND Corporation, TR-528-AF, 2008).

removed and sent to the Ogden Air Logistics Complex (OO-ALC). Later, repaired, refurbished, and/or upgraded versions of the engines and landing gear would be returned to WR-ALC for reinsertion on C-5 aircraft.[9]

Supply Chain Management Groups

As seen in Figure A.2, each depot has a supply chain management group responsible for requirements planning and executing the flow of supplies (e.g., broken parts, repaired parts, and new spare parts) into and out of the depot's maintenance facilities.[10] Supply chain management groups have topical ties to their depots (e.g., the supply personnel in Tinker Air Force Base's 848th Supply Chain Management Group generally manage parts used in aircraft engines).

Among the employees in depot supply chain management groups are item managers. As the title implies, an item manager has responsibility for a specific item (or NIIN) used on an Air Force system. In fact, individual item managers typically manage multiple, sometimes scores of, items at a time. The complexity of the item manager's job varies considerably by NIIN, so there tends to be an inverse relationship between the average complexity of an item manager's NIINs and the number of NIINs he or she manages. The item manager tracks where and on which end items the NIIN is in use in the Air Force (and possibly elsewhere [e.g., the Navy, foreign militaries]), the pattern by which it is failing, where the NIIN can be repaired and at what cost, and where and for what price new versions of the NIIN might be procured. The item manager will also monitor lead times (i.e., the typical time between when the item is put into repair and when it is repaired and the typical time between when an order is placed for a new item and when the item arrives from its manufacturer).

---

9 Some parts are "serial number tracked" to aircraft, i.e., the exact same part will be put back on the aircraft from which it came. More commonly, however, a different but functionally equivalent (or improved) part will be put on the aircraft as it completes PDM. The depots can cannibalize parts across aircraft to address parts shortages, e.g., remove a working radio from an incoming aircraft and insert it into an aircraft that is about to complete PDM.

10 See Mike W. Ray, "Alsup Named 448th SCMW Director," *Air Force Print News Today*, March 15, 2013.

The item manager also tracks the NIIN throughout its life cycle (e.g., whether it is to be replaced by a different NIIN or whether the end items on which it is used are being phased out). One of the most challenging scenarios for an item manager is when a NIIN is being phased out of usage. The item manager needs to continue to support customers to the end of a NIIN's life, but the item manager does not want to end up with a large and costly inventory of a NIIN that the Air Force no longer needs. A common approach is to leave a buffer of unserviceable, i.e., broken, NIINs in storage (with only a limited buffer of serviceable items) but only pay to repair the NIINs when the item manager receives a demand for the NIIN. The lengthier the NIIN's repair lead time, the larger the serviceable buffer is required to be.

DLR Planning

The Air Force aspires to have a "pull" inventory management system in which repairs are only undertaken on demand or when serviceable inventory levels fall below desired minimum levels rather than attempting to repair in front of demand by anticipating what might be needed. When the Air Force has reparable inventory levels above its current needs, that excess inventory tends to be kept in an unserviceable, rather than serviceable, form. Air Force item managers do not want to pay to repair an item until and unless they are fairly certain it will be needed. Figure A.4 illustrates this phenomenon for NIIN 008755009, a bulkhead hub mounting assembly used on a C-130 cargo aircraft propeller, managed by WR-ALC.

For this NIIN, the Air Force has modest serviceable inventory—sufficient to effectively serve customers—but sizable unserviceable inventory. When a demand arrives for this NIIN, repair must commence in short order in light of the limited serviceable inventory for this item.

The Air Force uses a system called EXPRESS to determine which items organic depots should repair when and which requisitions should

Figure A.4
NIIN 008755009 Illustrates Relatively Lean Serviceable Inventory

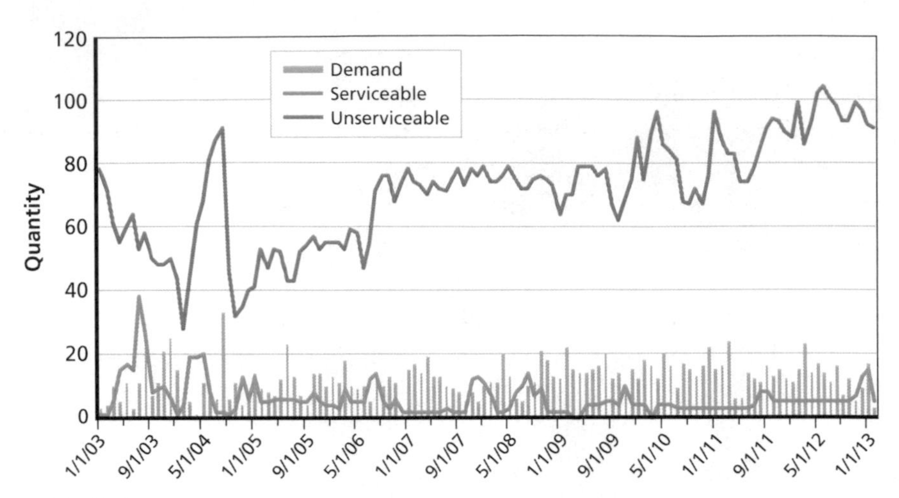

RAND *RR398-A.4*

be filled.[11] The goal of EXPRESS is to prioritize repairs to maximize aircraft readiness subject to constraints: capacity, funding, parts, and carcasses. EXPRESS uses information on Air Force readiness needs (e.g., which aircraft are grounded due to a lack of parts, referred to as a MICAP), demands, inventory levels, the presence of an unserviceable carcass to repair, depot capacity, funds availability, and parts availability to run a daily prioritization of which specific NIINs depots should put into their repair processes that day and which units should receive the repaired NIINs in serviceable inventory or coming out of production.

[11] More details on EXPRESS are provided in U.S. Air Force Materiel Command Instruction 23-120, *Material Management: Execution and Prioritization Repair Support System (EXPRESS)*, May 24, 2006. EXPRESS is descended from a RAND-developed algorithm called DRIVE (Distribution and Repair in Variable Environments), described in John B. Abell, L. W. Miller, Curtis E. Neumann, and Judith E. Payne, *DRIVE (Distribution and Repair in Variable Environments): Enhancing the Responsiveness of Depot Repair*, Santa Monica, Calif.: RAND Corporation, R-3888-AF, 1992, and L. W. Miller and John B. Abell, *DRIVE (Distribution and Repair in Variable Environments): Design and Operation of the Ogden Prototype*, Santa Monica, Calif.: RAND Corporation, R-4158-AF, 1992. EXPRESS does not apply to repairs purchased from contractors.

There is a cost-timeliness trade-off involved in depot maintenance workload management. A customer wants a part on demand. To have such responsiveness without having a large serviceable inventory buffer, a supplier like a depot would need to have a highly responsive, flexible repair system that can "turn on a dime," quickly changing from repairing one NIIN to repairing another, potentially very different, NIIN. From a depot's perspective, however, it would facilitate planning, management, and productivity if workload was planned months in advance. A maintenance depot can operate with the greatest efficiency when workload is smoothed over time, allowing full and stable usage of equipment, facilities, and labor.

The extent to which it is practical and cost-effective to switch depot maintenance and repair production processes varies by NIIN. For instance, an electronics test stand might be easily switched between different NIINs (with the turn of a dial, say). For that type of depot repair operation, a daily workload re-optimization in EXPRESS works well.

Rapid adjustment and sizable workload swings are less practical for a shop handling large, bulky items like wings, landing gears, or engines with long flow times and substantial input resource requirements. Shops may have limited space, and WIP of very large items introduces another constraint (floor space) that is not captured in EXPRESS's depot-capacity measure. For these types of items, daily demand-based inductions may not be the best overall solution; it may be more cost-effective to batch repair or maintenance workload. In fact, the actual implementation of depot maintenance on a number of items is not as flexible as implied by daily EXPRESS runs. Instead, the depots are using storage buffers, WIP, capacity (labeled "M") switches in EXPRESS, and other workarounds to regulate their work flow. These steps appear entirely reasonable and appropriate.

Interactions with the Defense Logistics Agency

The Air Force supply chain management groups are fundamentally integrated with DLA. In particular, DLA owns and provides requisite piece parts to depot maintenance personnel. Depot maintenance personnel order parts from DLA that are essential to their repairs.[12]

The 2005, BRAC called for DoD to reconfigure its industrial supply, storage, and distribution infrastructure so that one integrated provider, DLA, could support depot maintenance requirements. The argument was that this consolidation would reduce the duplication of functions and inventory, optimize resources, and streamline processes.[13]

With the transition, there have been parts support issues. But recently, there have been process improvements, e.g., the increased use of DHAs. With a DHA, DLA can factor in a demand it did not see into its forecasting (e.g., an Air Force maintainer using a back shop to produce a part that could not be obtained in a timely way from DLA). Without DHAs, DLA would not know its inventory level of a NIIN is likely inadequate because it would not have seen some of the demand for the NIIN. In a similar vein, a DDE is to be used to proactively notify DLA when there is a significant change in the future requirements for a DLA-managed consumable item. The 448th SCMW is responsible for transmitting DDEs to DLA.[14] Without the use of a DDE, DLA would learn of a demand change only retrospectively, as its demand forecasts gradually and reactively accounted for the change.

DHAs, DDEs, and other steps that might be implemented to reduce "seam problems" between military service maintenance personnel and the DLA supply system were further discussed in Chapter Three.

---

[12] Reparable items (i.e., items repaired by depots) are managed and funded by the Air Force but procured by DLA. Consumable items, by contrast, are managed by DLA. DLA has all supply, storage, and distribution responsibility for consumable items.

[13] See Amanda Creel, "Robins First Air Logistics Center to Implement BRAC 2005 Decisions," *Air Force Print News Today*, April 27, 2007. The U.S. Government Accountability Office has found that the level of cost savings from this consolidation will be smaller than initially anticipated (U.S. Government Accountability Office, "Military Base Realignments and Closures: Updated Costs and Savings Estimates from BRAC 2005," GAO-12-709R, June 29, 2012.

[14] See Headquarters, Air Force Materiel Command, Air Force Materiel Command Instruction 23-205, p. 8.

APPENDIX B

Army DLR Management

Organization

Among other missions, AMC is responsible for the item management of Army-managed DLRs. AMC reports on its activities to the Army Deputy Chief of Staff for Logistics (DCS G4), who provides Army logistics policy and process oversight on behalf of the Chief of Staff, and to the Assistant Secretary of the Army for Financial Management and Comptroller (ASA(FM&C)), who provides financial oversight and budget approval. Individual parts are managed by one of three subordinate commands, each of which workloads and manages repair depots associated with the command. This structure is shown in Figure B.1.[1]

- AMCOM—manages DLRs for aircraft and ground support equipment, as well as missile systems and two depot maintenance activities:
  - Corpus Christi Army Depot (CCAD), Corpus Christi, Texas—repairs and overhauls rotary-wing aircraft, engines, and components
  - Leterkenny Army Depot (LEAD), Letterkenny, Pennsylvania—repairs and overhauls missile systems

[1] Additionally, there are oversight organizations that monitor policy and procedures for depot maintenance; these organizations include the Depot Maintenance Corporate Board and the Army Working Capital Fund Requirements Review Group.

Figure B.1
Army Organizational Chart for Depot Repair

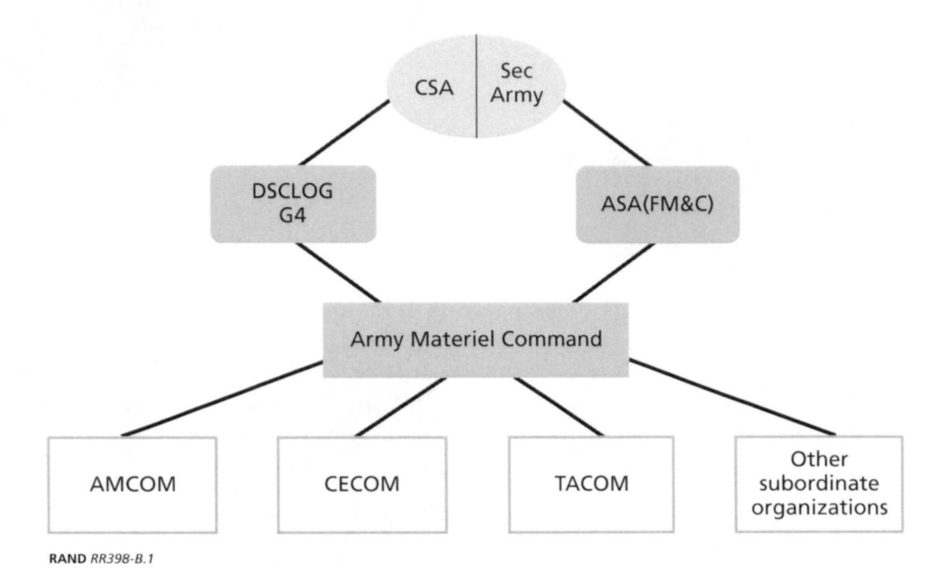

RAND *RR398-B.1*

- Communications and Electronics Life Cycle Management Command (CECOM)—manages DLRs for communications and electronics equipment and one depot maintenance activity:
  - Tobyhanna Army Depot (TYAD), Tobyhanna, Pennsylvania—repairs and tests communications-electronics equipment and repairs missile guidance systems
- TACOM—manages DLRs for soldier personal equipment, ground combat systems, support and construction vehicles and equipment, individual and crew-served weapons, and chemical and fire control equipment, and it manages two depot maintenance activities:
  - Anniston Army Depot (ANAD), Anniston, Alabama—repairs and overhauls tracked combat vehicles, self-propelled and towed artillery, generators, and rail equipment
  - Red River Army Depot, Red River, Texas—repairs light armored vehicles and some missile systems.

The locations of the commands and depots are shown in Figure B.2.

The Army also has a National Maintenance Program (NMP) through which it repairs selected DLRs at installation repair facilities also known as Directorates of Logistics (DOLs). DOLs are workloaded by the Life Cycle Management Commands (LCMCs) based on their capabilities, system capacity, and cost considerations.

Figure B.2
Army Repair Parts Management Commands and Maintenance Depots

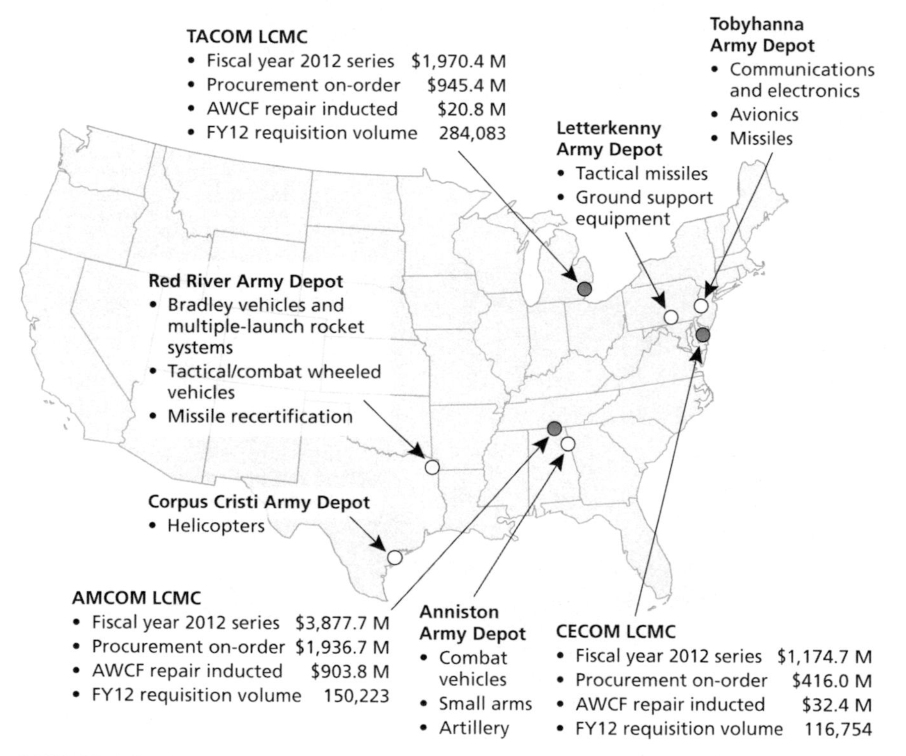

SOURCE: John T. LaFalce, "AMC Repair Parts Supply Chain," *Army Logistician*, May–June 2009.
NOTE: The numbers have been updated to FY 2012.
RAND *RR398-B.2*

Supply Planning

When new equipment comes into the Army or existing equipment is modernized, the program manager (PM) works with the equipment manufacturer to develop a list of consumable and repairable parts needed to keep the equipment operationally ready. Once soldiers begin using the equipment, item managers (IMs) assume the responsibility for providing and maintaining the inventory of parts to fill operational demands and provide parts for component repair. In addition to filling the demands of Army customers, requests for these parts from other services and foreign military sales also go through the LCMCs. Each IM is responsible for a group of parts usually unique to one weapon system. The IM directs the issue, repair, and procurement of parts to maintain an overall stock availability goal. The aggregate goal is 85 percent, which is allocated by item groups and then within groups based on minimizing the safety stock investment to achieve the target for each group.

If unserviceable inventory of DLRs is available, IMs first turn to the repair of broken but economically reparable components rather than buying new components. This can be impeded, though, by repair capacity or parts support, as well as insufficient carcass inventory, necessitating buys for effective customer support. All buy requirements greater than $2.5 million are reviewed by the IM's team lead, requirements officer, supervisor, group chief, and then by the tactical requirements officer and director. Purchases greater than $10 million go to the LCMC commanding general for approval.

Maintenance Composition

The five Army depots perform component repair and the recapitalization/reset of end items for each of their managing LCMCs. The depots produce DLRs for return to stock, as well as repair DLRs in direct support of end-item programs. Data for CCAD and ANAD illustrate this workload split.

For CCAD, workload has been steady at 66 percent DLRs and 34 percent airframes for several years. Seventy-five percent of the DLR workload, including engines, is funded through the Army Working Capital Fund (AWCF), producing components to replenish inventory or directly fill back orders. Most of the remaining component funding is from Operations and Maintenance Army (OMA) in support of end-item programs.

For ANAD, workload is split between vehicles/artillery (55 percent) and small arms/subassemblies/other (45 percent), with DLRs being among the small arms/subassemblies/other portion. Of the 45 percent of the work that is small arms/subassemblies/other, 39 percent is AWCF funded and 61 percent is OMA funded. So, in total, 18 percent of the workload is AWCF funded.

Information Systems

Field operating units place requests for parts through the Standard Army Retail Supply System (SARSS). The Army is currently replacing SARSS with Global Combat Support System–Army (GCSS-A). GCSS-A is a SAP-based automated logistics ERP system.

Orders from SARSS/GCSS-A for Army-managed items are routed to AMC's ERP system—the Logistics Modernization Program (LMP). CECOM was the first LCMC to begin using LMP in 2003; AMCOM began using it in 2009, and TACOM began in 2010. From the first to the fifteenth day of every month, item managers at each of the Army LCMCs run demand planning (DP) in LMP to update the forecasts for their items. The output of DP feeds material requirements planning (MRP), which is run on the 15th of the month. The output of MRP is a time-phased buy and repair induction plan based on projected demands, expected receipts from suppliers, expected completed repairs, forecasted condemnation rates, and procurement and repair lead times.

LMP feeds the information system run at Army depots known as the Army Workload and Performance System (AWPS).[2] AWPS has information on available skills, labor expenditures, performance data, and work schedules.[3] AWPS is updated daily and used for workforce planning. It is used to track core requirements by weapon system, depot or arsenal, workforce skills, weapon system quantities, direct labor hours of annual depot repair at organic depots, program cost, and schedule performance. Data from AWPS feed into the Operational Program Summary (OPS)-29 depot maintenance requirements determination and programming process each budget cycle.

Annual Production Planning and Depot Workloading

Each FY, production planning and depot workloading begins with the development of the Depot Maintenance Requirements Plan (OPS-29) for the following year. About nine months in advance of the FY, planners begin work on the command schedule for the following year. From March through August, the monthly MRP outputs are reviewed to ensure the accuracy of projections and input any updated and improved data.[4]

Annual Budget Planning

Requirements identification is the first step in budgeting. U.S. Army Regulation 800-90, *Army Industrial Base Process*, describes the annual repair and budget planning process. Each FY's planning process begins with producing the depot maintenance requirements plan (OPS-29).

---

[2] All Army depots use AWPS for reporting purposes. Some Army depots also use AWPS for depot management; those depots not using AWPS have local models tailored to their specific needs.

[3] "The Army began developing AWPS in 1996 at the direction of the House National Security Committee (now the House Armed Services Committee)." See U.S. Government Accountability Office, *Oversight and a Coordinated Strategy Needed to Implement the Army Workload and Performance System*, GAO-11-566R, July 14, 2011, p. 9.

[4] The Army's annual repair and budget planning is described in U.S. Army Regulation 700-90.

During the OPS-29 review process, core workload requirements are validated as critical, which enables the identification of funding levels needed to sustain minimum core capabilities within the base program.[5]

It is the job of the Army Organic Industrial Base (AOIB) to identify, prioritize, and resource (within funding constraints) depot workload.

> The AOIB has made a significant change in the way core depot requirements are viewed and prioritized in the POM [Program Objective Memorandum] or FYDP [Future Years Defense Plan] budget development process. The change includes highlighting and prioritizing core requirements that are met from among the various depot maintenance requirements. These requirements include depot cyclic overhaul; the demand driven secondary item AWCF depot-level reparable (DLR) component repair program; Reset and Recapitalization program depot repair, overhaul or rebuild; requirements identified through the Aircraft and Combat Vehicle Evaluation (ACE/CVE) Teams; and requirements associated with support for modified, upgraded, and new weapon systems.[6]

Planning the repair schedule begins in December and January for the next FY, with IMs using forecasts from the MRP module of LMP. End-item recapitalization and reset[7] requirements are loaded into LMP before March. The Department of the Army and PMs validate flying hours (air systems), miles driven (wheel systems), and hours used (track

---

5 "Depot maintenance requirements are reviewed in the Operational Program Summary (OPS)-29 Review Process utilizing the Depot Maintenance Operations Planning System (DMOPS). Key players in the depot maintenance requirements and budgeting processes include the HQAMC LCMCs, Program Executive Officers (PEOs)/PMs, ARNG [Army National Guard], and USAR [U.S. Army]" (U.S. Army, *Organic Industrial Base Strategic Plan (AOIBSP) 2012–2022*, undated, p. 14).

6 U.S. Army, undated.

7 *Recapitalization* refers to the complete rebuilding and selected upgrading of currently fielded equipment to new condition with zero time/miles. *Reset* refers to restoring equipment to optimal capability after redeployment from a combat or stability operation.

Table B.1
Budget-Year Time Line

Sequence Number	Tasks	Subtasks	Action	Start Date	Finish Date	Remarks
1	Develop the NMP Repair Requirements forecast using MRP and other data available through LMP.	Develop Requirements Forecast IAW AMC G4 OPS29/G8 POM Guidance	LCMCs	November	November 30	
2	Input provided for POM process.	LCMCs input Below Depot Requirements in to LMP Project System. This data is passed to DMOPS.	NLCO LCMCs	December	No later than January 5	
3	Budget-Year data extracted	LCMCs may validate budget-year data in DMOPS.	LCMCs	May 1	No later than May 15	Data available after May 1
4	Draft Budget-Year Workload Plan	AMC NLCO pulls and forwards consolidated plan to NSFLMD	AMC NLCO	May 15	May 20	
5	Budget-Year Plan Published	AMC G4 publish approved Budget-Year Plan to ACOM/ASCC/ARNG (24-month notice)	NSFLMD	May 20	June 1	

SOURCE: U.S. Army Materiel Command, *National Maintenance Program Business Process Manual*, Chapter 4: National Workload Plan Development and Validation, October 1, 2012.

NOTE: ACOM = Army Command, ASCC = Army Service Component Command, NLCO = National Logistics Coordination Office, NSFLMD = National Sustainment and Field Level Maintenance Division.

Table B.2
Apportionment Year–Validation Time Line

Sequence Number	Tasks	Subtasks	Action	Calendar Days	Start Date	Finish Date	Remarks
1	Plan development strategy and publish repair requirement projection and SOR selection guidance.	1. Prepare and staff HQAMC guidance memo 2. Prepare enclosures to memo (if required) 3. Publish	AMC G4, AMC NLCO, and LCMCs	30	Jan 1	Jan 30	Provide format for all LCMCs' workload submissions. Issue workload assignment guidance to facilitate LCMC SOR selection recommendations.
2	Request labor rate data call from ACOM/ASCC/ARNG.	ACOM/ASCC/ARNG identify labor rates.	AMC G4, ACOM/ASCC/ARNG.	45	Jan 10	Feb 25	
3	Labor rates provided to LCMCs.	Labor rates are loaded by LMP NMP Role 20, AMC—Maintenance Business Team Lead (BTL)—NMP BTL Administrator	AMC G4, LCMCs and Domain Manager	10	April 1	April 11	
4	LCMC update projects based on validation process	LCMC insures audit trail of change process is maintained	LCMCs	13	April 12	April 25	

Table B.2—Continued

Sequence Number	Tasks	Subtasks	Action	Calendar Days	Start Date	Finish Date	Remarks
5	LCMC sends out-year report to SORs LCMC creates and processes CIR as necessary	LCMCs coordinate repair programs with SORs. SORs provide cost and capability information to LCMCs.	LCMCs and SORs	21	May 15	June 6	Notify SOR by email of any changes after LCMC transmits the initial out-year report.

SOURCE: U.S. Army Materiel Command, 2012.

NOTE: SOR = source of repair.

systems) by weapon system.[8] In March, the requirements are "locked down" and the first budget submission is made. This initiates a back-and-forth discussion on assumptions with IMs, PMs, AMC G4, and the Army Budget Office (ABO) to finalize the budget. Detailed information for planning for the budget year (current year + 2) and validation of the apportionment year (current year + 1) is provided in Tables B.1 and B.2.

Program change factors (PCFs) are used in LMP to change demand forecasts that will be impacted by drawdown or surge plans or changes in fleet sizes. PCFs allow changes to be entered at the weapon system/major item level, with changes in plans then cascading to all of the indentured parts. PCFs require production bills of material (PBOMs), which contain all parts for a system. However, with LMP's recent fielding, the LCMCs have first concentrated on building and fine-tuning repair bills of materials (RBOMs), which contain only the items in a PBOM needed for repair, because repair is the preferred method of supply. Because there are few PBOMs, program change factors are manually put into LMP at the component level for all items used on a weapon system/major item.

Allocation of Funding

Personnel at headquarters AMC reported that they provide 75 percent of hardware (repairs and procurements) funding up front and 50 percent of logistics operations (e.g., to fund LCMC personnel) up front. Funding is allocated across the LCMCs as part of the budget process, and there is no reallocation during the year unless there is a big change in requirements. Each LCMC manages its AWCF funding, employing somewhat different approaches.

AMCOM fully funds all repairs for the year up front. Most TACOM programs are funded at 50 percent up front. After the first quarter of the year, there are reviews for programs and the second half

[8] HQ AMC requires that major item planning books are updated annually—typically in September—to ensure new FY projections are correct. DA G4 and PMs meet semiannually to review OPTEMPO projections. Validation of previously submitted OPTEMPO data occurs quarterly.

of the funding is typically provided in the second quarter, after the January review. Depot repair contracts can be cut if demands go down; however, the main reason for not completing a repair contract is a lack of unserviceable carcasses for repair. CECOM has recently switched from funding 100 percent of its programs at the beginning of the year to incremental funding.

Workload Adjustment

According to AMC personnel, the command schedule for repair is agreed upon before the beginning of the FY and never changes. The actual timing of repair execution within the year results from negotiation between depots and IMs, and the actual depot production schedule is changed during the year if necessary based on depot needs and changes in demand. Production schedule meetings are held at the end of each month to focus on repair programs that are behind schedule or over cost. The schedule is adjusted on the first day of the following month. Depots hold quarterly reviews of all repair programs with their managing LCMC. If the updated forecast produces a need for a 15 percent change or more, then the repair schedule may be adjusted.

If an increase is required, repair capacity must be checked for adequacy, with the LCMC Industrial Base Operations office involved in the decision to increase the program and the coordination to determine if there is sufficient capacity. In recent years, funding has not been a constraint on making needed changes. As an example of the process, when asked to adjust or increase workload, CCAD performs an analysis, checking for the availability of parts, labor, tools, and equipment, to determine the feasibility and effects of the change. This analysis assesses the probability of success and identifies high-risk areas and long lead items.

The repair contracts for NIINs that have two sources of repair are used to handle surge capacity and variability in demand, with the depots used to handle the more stable portion of demand. For systems such as the Blackhawk and Kiowa, which have both organic and contractor repair, the first choice is generally to change contractor workload rather than depot workload, but the contracts must be written to allow for such changes. This saves depot capacity for flexibility to repair different items or handle changes for items for which the depot is the sole source. The advantage of organic maintenance is that the LCMC can tell the depot to change the schedule and work on something else, without contract constraints, given that the right resources are available. That flexibility is not possible in the short term with contractors if a contract is not in place or a contract is not designed with that flexibility. All interviewees stated that organic depots are more flexible with regard to changing workload than commercial repair.

For all depots, demands may diverge from forecasts due to changes in field OPTEMPO or deployment plans or simply from stochastic variability, which can have a larger impact on lower-demand DLRs. In addition, there are other factors that can influence forecasts. Most major changes in DLR forecasts are due to changes in repair plans for end items, such as changes to recapitalization or reset programs. When plans change, PMs will call with the information on changes. In addition to major end-item program changes, safety of flight issues or other part upgrades can cause program changes for a specific part.

Process for Providing Planning Information to DLA

When production continues as planned, the depots use the following planning lead times for ordering DLA parts, based on the Acquisition Advice Code (AAC):

- D-45 days (stocked items)
- Z-75 days; these parts are called "insurance items" (stocked but not demand supported with a potentially higher risk of not being available if the forecast is high)
- J-120 days; these parts are not stocked and have long lead times.

When program changes are considered, to help DLA prepare for the shift, the IM uses special program requests (SPRs) for the parts forecasted for DLA. To check on the availability of needed parts, IMs run the forecast availability module in DLA's Electronic Mall (EMALL). They email Weapon System Support Managers (WSSMs) at DLA or the LCMC liaison to verify the availability information for the parts they need and to inform them of program changes. In addition, depot and DLA personnel have weekly meetings to get in front of problems, with DLA providing estimated ship dates for outstanding orders.

For coordination with DLA, the limited number of accurate, available BOMs is a problem. Many PBOMs are not accurate and only about half have been built (TACOM). Even the high mobility multipurpose wheeled vehicle (HMMWV) does not have a complete PBOM because it is difficult to build to LMP accuracy requirements. Because there are so few accurate PBOMs, communication with DLA cannot be automated. Manual intervention is required to assure parts support. Depots are responsible for RBOMs, but IMs, depots, and DLA work together to improve the RBOMs.

APPENDIX C

Navy DLR Management

Organization

Supply Planning

Item management for DLRs, including supply planning, falls under the jurisdiction of the Navy Supply Systems Command (NAVSUP). NAVSUP consists of four activities: NAVSUP Weapon Systems Support (WSS), NAVSUP Business Systems Center, NAVSUP Exchange Service Command, and NAVSUP Global Logistics Support. NAVSUP WSS has primary responsibility for DLR item management.

Previously known as the Navy Inventory Control Point (NAVICP), NAVSUP WSS was formed in 2011. Created in 1995, NAVICP consolidated all of the Navy's supply planning and management functions under a single command, uniting the Aviation Supply Office (ASO) in Philadelphia, Pennsylvania, and the Ships Parts Control Center (SPCC) in Mechanicsburg, Pennsylvania. NAVSUP WSS headquarters is also located in Mechanicsburg, Pennsylvania.

Item managers at NAVSUP WSS play an important role in monitoring item inventory levels and assessing the ability to meet customer demand. Using inventory management software, item managers establish target inventory stock levels for each item, as well as target levels for inventory positioned at each operational and maintenance site. To maintain target inventory levels, item managers determine when to induct items for repair and when to procure new items. The item manager decides where to send the inducted carcass for repair and routes the carcass to the appropriate location. When new DLRs need to be procured from commercial suppliers, NAVSUP WSS works with DLA, which executes the procurement action. NAVSUP contracts directly with commercial providers for DLR repair services.

Maintenance Structure

In contrast with the Navy's supply planning organization, which is centralized within one systems command across the Navy, the maintenance of aviation, ship, and communications and information systems DLRs is managed by three different systems commands. The next two subsections review how the aviation and ship maintenance functions are organized. We did not include communications and information system DLRs in our analysis. These items are maintained by the Space and Naval Warfare Systems Command.

Aviation Maintenance

Aviation maintenance falls under the Naval Air Systems Command (NAVAIR). As directed by the 2005 BRAC legislation, the Navy consolidated three primary Navy aviation depots and 18 intermediate maintenance facilities into eight fleet readiness centers (FRCs), as depicted in Figure C.1. These FRCs include

- FRC Southwest (Coronado, California)
- FRC Southeast (Jacksonville, Florida)
- FRC East (Cherry Point, North Carolina)
- FRC Mid-Atlantic (Virginia Beach, Virginia)
- FRC West (Lemoore, California)
- FRC Northwest (Whidbey Island, Washington)
- FRC WestPac (Astugi, Japan)
- FRC Support Equipment Facility (Solomons Island, Maryland).

Command of the FRCs belongs to Commander, Fleet Readiness Centers (COMFRC), headquartered in Patuxent River, Maryland. COMFRC also has responsibility for managing maintenance and repair performed by private contractors. Nearly 18,000 personnel, of whom roughly two-thirds are civilians or contractors, fall under the command of COMFRC. FRCs that focus on intermediate-level maintenance tend to have a greater share of military personnel; by contrast,

Figure C.1
FRC Locations and Workload

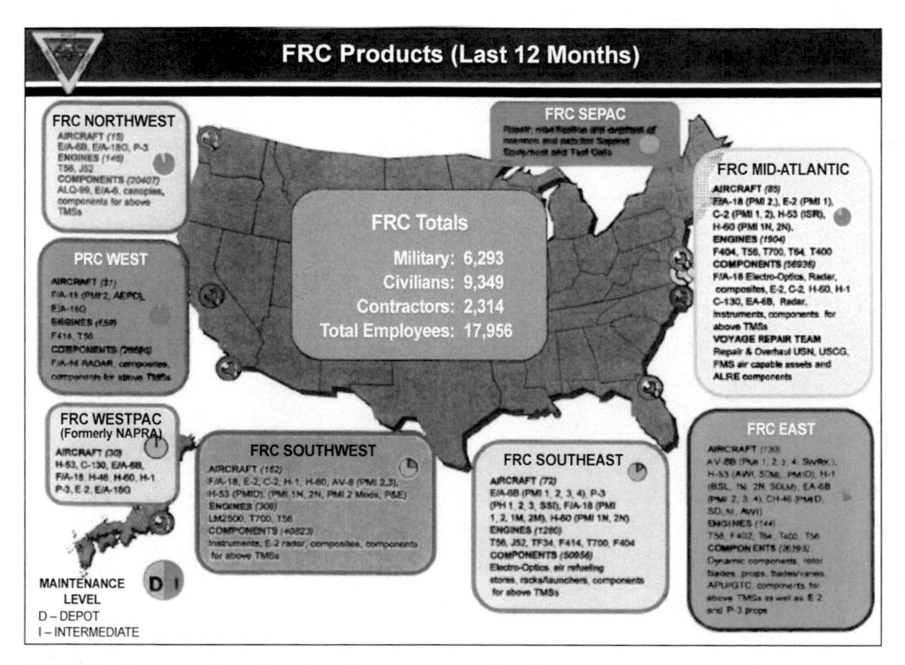

SOURCE: Commander, Fleet Readiness Centers briefing.
RAND *RR398-C.1*

civilians and contractors make up a larger share at FRCs with a focus on depot-level work.

As indicated above, there are multiple levels of Navy aviation maintenance, including base/organizational-, intermediate-, and depot-level maintenance. When a DLR fails, maintenance personnel belonging to the operational command will attempt to repair the item. Detachments from FRCs (called Voyage Repair Teams) are forward-deployed to complement base repair efforts, providing some intermediate maintenance and repair capabilities on aircraft carriers and amphibious ships; smaller ships, such as guided missile destroyers (DDGs), do not have intermediate maintenance capabilities during deployment. Aircraft carriers and amphibious ships maintain pools of spare parts for aircraft, as specified in the Aviation Consolidated Allowance List (AVCAL). While the other services rely on the other echelons or reachback for maintenance beyond the organizational/direct-support level, the Navy deploys with intermediate maintenance capability. This difference in strategy is rooted in practical considerations: the Navy deploys its bases (i.e., ships), whereas aircraft in the other services move from one base to another.

If a repair is too complex to be handled at the organizational level, the failed item is sent back to the Aviation Support Division (ASD) and a replacement item is ordered. If a replacement item is available, it will be issued; otherwise, it is back-ordered. Occasionally, IMs may hold available items for higher-priority demands in reserve.

When failed items cannot be repaired at the organizational level, ASD reviews the Individual Component Repair List (ICRL) to determine whether an item can be repaired at the intermediate level. If it can, ASD sends the item to the appropriate intermediate-level site. The key FRCs focusing on intermediate-level maintenance include FRC Mid-Atlantic, FRC Northwest, and FRC West. Once the item arrives at the intermediate-level site, the FRC inspects the item and evaluates whether it has the capability to repair the item. Intermediate-level maintenance personnel are complemented by depot-level artisans, providing added capability to repair an item at the intermediate level. If capable, personnel at the FRC repair the item. However, if they lack the capability, the item is sent back to ASD, which then sends the item to a depot for repair. ASD also sends items not on the ICRL to the depot for repair. The primary depot-level maintenance functions occur at FRC Southwest, FRC Southeast, and FRC East.

Ship Maintenance

Analogous to NAVAIR command of aircraft maintenance, control of ship maintenance falls under the Naval Sea Systems Command (NAVSEA). However, the organization and procedures for repairing ship DLRs diverge from those in place for repairing aircraft DLRs. While FRCs maintain primary responsibility for intermediate- and depot-level maintenance, responsibility for the ship maintenance is divided between the four public shipyards and the six regional maintenance centers (RMCs), with some of the warfare centers, including Crane, Carderock, and Newport, also playing a small role in ship maintenance—albeit a more significant role with respect to the focus of this report, DLRs. The four public shipyards include Norfolk Naval Shipyard in Portsmouth, Virginia; Portsmouth Naval Shipyard in Kittery, Maine; Puget Naval Shipyard and Intermediate Maintenance Facility in Bremerton, Washington; and Pearl Harbor Naval Shipyard and Intermediate Maintenance Facility in Pearl Harbor, Hawaii. The locations of the original eight RMCs are depicted in Figure C.2. The shipyards and RMCs have distinct command reporting structures, with the shipyards reporting to the Logistics, Maintenance, and Industrial Operations Directorate (SEA04) and the RMCs reporting to Commander Navy Regional Maintenance Center (CNRMC); both SEA04 and CNRMC report to the NAVSEA commanding officer. Table C.1 summarizes the locations of the shipyards and RMCs.

Similar to aircraft maintenance, there are three levels of ship maintenance: organizational/base-, intermediate-, and depot-level maintenance. Repair personnel first try to repair items at the organizational level. Aircraft carriers and amphibious ships carry pools of spares, as specified in the Consolidated Ship Shipboard Allowance List (COSAL). Failed items requiring intermediate-level repair are sent to either shipyards or RMCs, depending on the needed capabilities. The RMCs focus on surface ship maintenance and also perform non-nuclear carrier repairs. Nuclear items are repaired exclusively at the shipyards, with submarine items repaired both at the shipyards and at warfare centers that specialize in undersea DLRs. Ship and undersea DLRs with faults beyond the repair capabilities of intermediate-level facilities are sent to depot-level facilities, which are located at the shipyards and warfare centers. The majority of depot-level ship repair takes place during ship availabilities (periods when ships are undergoing scheduled maintenance and not available for operations); by contrast, aircraft DLRs are removed and shipped for repair, with a serviceable replacement DLR provided in exchange, in higher proportions. In addition to ship availability taskings, shipyards receive some taskings from NAVSUP, as do warfare centers.

Note that RMCs differ considerably from FRCs. In particular, RMCs do not perform depot-level work. Furthermore, repairs made by RMCs are returned directly to the customer; RMCs do not perform

Figure C.2
Eight Original RMC Locations

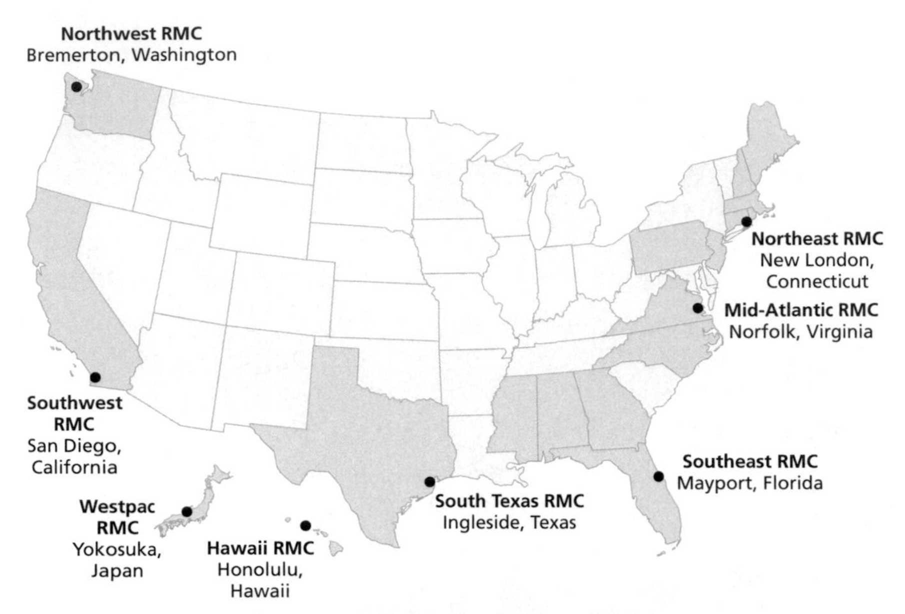

SOURCE: U.S. Government Accountability Office, "Navy Regional Maintenance: Substantial Opportunities Exist to Build on Infrastructure Streamlining Progress," 1997.
NOTE: South Texas RMC and Northeast RMC no longer exist. Mid-Atlantic RMC has been renamed Norfolk Ship Support Activity (NSSA).
RAND *RR398-C.2*

ready for issue (RFI) work. In other words, RMCs perform "repair and return" taskings, as opposed to the "repair and stock" taskings that take place at intermediate-level FRCs. In addition to intermediate-level maintenance, RMCs perform several other functions. Training is a particularly key component of the work portfolio of the RMCs, preparing sailors to perform maintenance and provide shipboard support while on deployment. Contracting with OEMs and private repair firms, as well as overseeing their operations, represents another key function of the RMCs; ship maintenance tends to rely more on private contractor repair than aircraft maintenance. Engineers at the RMCs also provide technical assistance to the fleet. RMCs consolidate functions performed at the former ship intermediate maintenance facilities

(SIMAs), supervisors of shipbuilding (SUPSHIPs), readiness support groups (RSGs), and fleet technical support centers (FTSCs).

Information Systems Overview

In 2010, the Navy started implementing an ERP system, replacing the legacy Uniform Inventory Control Program (UICP) system. Anchored by SAP's ERP software, the Navy's ERP program consists of several software packages. The Readiness Suite is used for demand-based item level setting for fleet and shore activities. In addition, the ERP contains a readiness-based sparing add-on package based on Morris Cohen Associates Supply Planning and Optimization (MCA SPO) that is used by the Navy to help set spare requirements. These systems are integrated with the base SAP system with the input data described below for multi-echelon inventory level setting. Note that the inventory modeling software provides recommended sparing level requirements; IMs then review the model's recommendations and override or adjust the recommendations as necessary.

One of the key motivations for implementing the ERP system was to centralize and simplify the management of items for IMs at NAVSUP WSS. All aspects of item management, including financial and operational considerations, are integrated into ERP. Furthermore, ERP helps to provide total asset visibility across the nodes of the repair and maintenance system. Using ERP, NAVSUP WSS has visibility of deployed ship inventories, including AVCALs and COSALs.

The ERP incorporates a wide breadth of item input data into its model. One key input is the item failure rate based on engineering estimates or experience. For new items, engineering estimates of the failure rate, known as the technical replacement factor (TRF), are provided by OEMs. The TRF is specified as the number of expected item failures per year. After the item has been deployed for 18 months, the IM develops a best replacement factor (BRF), taking into account the observed failure rate. For older items, there may be ample historical data; however, experienced failure rates could be influenced by wartime usage levels. The model also incorporates repair lead times and requirements, as well as acquisition lead time and cost data, to help make inventory decisions and determine optimal stock levels. Data quality is critical for ERP performance. In particular, forecasting demand represents one of the most significant challenges for IMs. Logistics personnel at NAVSEA and NAVAIR help NAVSUP IMs update demand and other item input data, a process conducted quarterly.

Push Versus Pull System Characterization

In general, the Navy's inventory system policy would be characterized as a pull system, although certain system nodes exhibit push-like qualities. Depot-level facilities are given eight-quarter workload plans, which are updated every quarter by item managers at NAVSUP. At the start of the FY, NAVSUP provides a six-month induction execution plan. For aviation DLRs, NAVSUP WSS Philadelphia then provides three-month plans just prior to the start of the 3rd and 4th quarters, with NAVSUP WSS Mechanicsburg providing a six-month plan for the second half of the year for ship DLRs. Item managers typically do not adjust the workload plan within these periods but can do so by exception to meet critical customer needs, creating emergent repair requirements that are passed to repair activities in such situations. Note that NAVSUP provides a semiannual or quarterly "induction plan," depending on the WSS, representing items to be inducted into repair, not a "production plan," representing what is to be completed in the period. Additionally, depot-level facilities have the freedom to schedule work during the workloaded period (i.e., three or six months) so long as they meet the overall induction requirement with items delivered within the standard or planning time frame for each, as reflected in the ERP input data used in the development of the plan. With this local control, depots are able to create a more "level" schedule for repairs that balances workload across the period, which is critical for repair facility efficiency. In addition, repair facilities batch repairs of DLRs, particularly items with long setup times. A lack of carcasses and the prioritization of repairs may limit the extent to which depots can batch repairs.

Typically, a first in, first out (FIFO) process is used to schedule work, although depots may deviate for high-priority customer demands.

By contrast, intermediate-level repair facilities are significantly more responsive to demand, in line with a pure pull system. Unlike depot-level facilities, intermediate-level facilities do not level-load, repairing on demand instead. Repair work is not batched and customer demand is satisfied according to priority. For intermediate-level facilities, flexibility and the ability to respond quickly to demand are the key objectives, not repair efficiency.

Funding

One of the key discriminators between intermediate-level and depot-level work, as well as ship and aviation repair, is the process for allocating funding. Historically, depot-level repair was funded through the Navy Working Capital Fund (NWCF). Under this system, aircraft and ship maintenance was managed by NAVAIR and NAVSEA, respectively. When customers at the fleets required depot-level services, they had to pay for the repairs using their allocated funds.

Under the Navy's Regional Maintenance Plan, the funding mechanism for shipyard depot-level maintenance was switched from the NWCF to primarily mission funding (ship maintenance facilities still receive some taskings from NAVSUP, which are NWCF-funded). Warfare center repair facilities remain NWCF funded. Under mission funding, the numbered fleets directly allocate a portion of their budgets to the shipyards to fund depot-level services; in contrast with the NWCF mechanism, mission funding does not require customers to pay for depot-level maintenance since they have already "purchased" services by allocating funds to the shipyards. In making the switch from NWCF to mission funding, the Navy gave several justifications. First, it stated that having common funding mechanisms for intermediate-level maintenance (which previously used mission funding) and depot-level maintenance would enhance the integration and consolidation of colocated intermediate- and depot-level facilities. In addition, the Navy emphasized the enhanced flexibility to redeploy resources, particularly labor, according to the fleets' priorities under mission funding. Finally, since most shipyard work is mission-funded and driven by ship availabilities, specific level-loading of relatively small NWCF-funded NAVSUP taskings is not necessary; such taskings can be easily integrated into the flow of ship availability tasking.

Although ship maintenance is generally mission-funded (with some repair at shipyards and all repairs at warfare centers NWCF-funded), aircraft maintenance continues to have separate funding mechanisms for intermediate- and depot-level maintenance. While intermediate-level aircraft facilities are mission funded, aircrafts depots are funded through NWCF. Since customers have to pay for services at the depot level, there is a strong incentive to have items repaired at the intermediate level, which is less costly. The integration of depot-level artisans at the intermediate-level facilities increases intermediate capabilities but makes financial tracking more complex, since the two labor pools are funded by "different colors of money."

Retail Parts Management and the Role of DLA

DLA plays an important role in the management and procurement of retail parts for the Navy. As part of BRAC 2005, over the last several years, the Navy has been migrating key aspects of retail parts management to DLA, although NAVSUP WSS retains certain responsibilities. While NAVSUP WSS determines the procurement requirements for reparables and works directly with internal Navy providers for any items manufactured in house, DLA has responsibility for purchasing DLRs from commercial suppliers. DLA also determines the procurement requirements for consumables and has taken over warehouse management of these items in support of depot maintenance, with the Navy responsible for providing information on changing demands for DLA to use to adjust demand forecasts based on historical data. While DLA representatives are embedded at depot-level facilities to facilitate this flow of information and respond to critical needs, they are not usually present at intermediate-level facilities. Finally, DLA manages the storage of items in warehouses, but the Navy retains ownership of the warehouse inventory.

APPENDIX D

Logic to Identify DLRs

We categorized an item as DLR if any of the following were identified for the NIIN. Multiple criteria were used because the services use NIIN identifiers and information systems in different ways.

The first criterion was Maintenance Repair Code. This code, primarily used by the Army, denotes the lowest level of maintenance authorized to conduct complete repairs:

- D (depot)
- L (specialized repair activity)
- H (general support level).[1]

When the Maintenance Repair Code was blank, we used the Recoverability Code (RC), if available. Again, this code is primarily used by the Army. The RC denotes the lowest level of maintenance authorized to determine if an item is uneconomically repairable and should be condemned and disposed:

- D (depot)
- L (specialized repair activity)
- H (general support level).

[1] While H items have not historically been considered DLRs, given the Army's adoption of the National Maintenance Program, in which the LCMCs workload DOL general support-level maintenance activities as part of depot maintenance planning, and its adoption of two-level maintenance, we included H items in the DLR category. We also included L items, which have been traditionally considered DLRs, due to the high level of depot-like repair capability found at specialized repair activities.

For the Air Force, we used the Expendability, Recoverability, Reparability Code (ERRC), with depot level indicated by C, S, or T (mostly T). And for the Navy, we used the Cognizance (COG) Symbol code or the Materiel Control Code (MCC), with the following indicating a DLR:

- COG: 7 series, 2F, 2J, or 2S
- MCC: of E, H, G, Q, or X.

For all services, regardless of the codes above, if a NIIN was seen to have substantial depot repairs conducted as reflected by D6Ms (receipts) from depot maintenance activities, we coded it as a DLR. This required the preponderance (greater than 50 percent) of receipts into DLA DCs to be from depot maintenance, with at least eight D6Ms from depot maintenance per year, on average.

APPENDIX E

Categorization of DLRs

To enhance our understanding of inventory management performance across the services, we developed an algorithm to categorize items. At any given moment, some items are in the phase-out stage of their life cycles, either because they have been replaced by upgrade DLRs or the associated end-item is being phased out; other items may be in the nascent phase of their life cycle, while the remaining items are in the middle of their life cycle. Key aggregate measures of performance, including inventory turnover and cost, are influenced by the life cycle position of individual DLRs. For instance, an organization with a large number of items that are being (or already have been) phased out of service may appear to have low inventory turnover, as inventory accumulates with a reduction in demand. By contrast, services with a significant number of newly deployed items with increasing demand may have higher turnover, particularly if the initial provisioning levels were stingy. Moreover, it is the steady-state phase of an item's life cycle that gives a clear picture of the level of assets needed to provide effective customer support. Thus, this period of an item's life cycle is the best to use to measure inventory turns in order to gauge aggregate supply chain process performance.

To overcome these challenges in evaluating inventory management performance, we devised a statistically based algorithm to categorize items based on empirical demand patterns. The algorithm applies heuristics and was developed iteratively using trial and error. We started with rules that we thought would isolate the categories, executed the algorithm, and then made adjustments to the rules until the targeted accuracy was reached. Given that the services each have tens of thousands of different items, an automated categorization approach is essential. This appendix explicates the algorithm and then describes the development process in more detail.

Algorithm Description

In the algorithm, we divide items into four categories: Steady State, Phase-In, Phase-Out, and Other. Categorizations are made with respect to a reference year, which we define as 2010. Our available dataset spans 2003 to 2012. Sufficient data need to be available before and after the reference year to determine how the demands in that year compare to the long-term trend, which is part of the categorization criteria. Qualitative definitions of these four categories are as follows:

1. *Steady State*: Steady-state items are in the middle of their service lives and do not have a discernible upward or downward demand trend.
2. *Phase-Out*: Phase-out items either went out of service before the reference year or are going out of service during the reference year, as exhibited by a downward demand trend.
3. *Phase-In*: Phase-in items are being phased into service during the reference year and have an upward demand trend.
4. *Other*: These items cannot be placed into any of the first three categories since there are insufficient data to estimate a time trend. In particular, these items have low and/or sporadic demand. The majority of these items have no items demanded over the 10-year dataset, indicating that they may have been phased out prior to 2003, the first year in our dataset. Even items with very low, sporadic demands may represent items that were phased out prior to 2003.

In Figure E.1, we depict typical examples of 10-year demand patterns for items that belong to each category. The chart in the top left-hand corner illustrates a steady-state item with no obvious upward or

Figure E.1
Examples of Different NIIN Categories

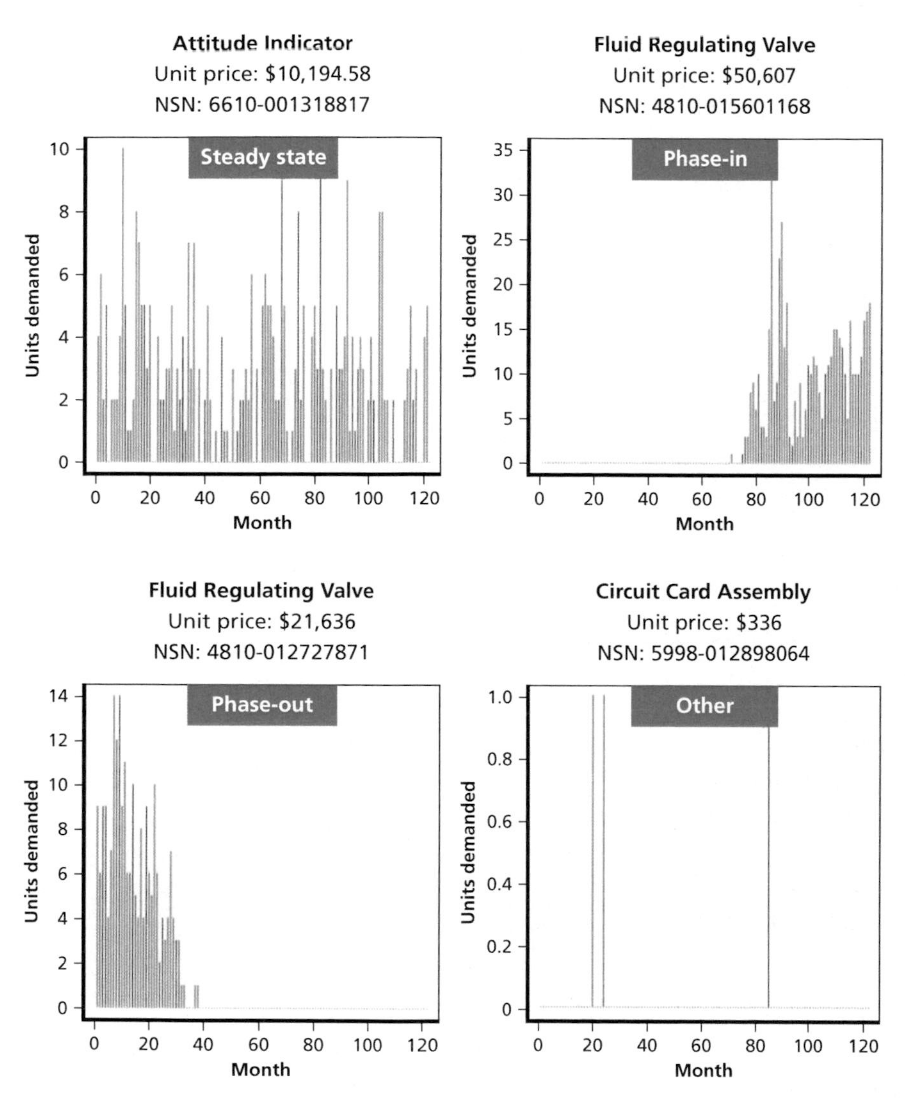

RAND *RR398-E.1*

downward demand trend. By contrast, the chart in the top right-hand corner depicts a phase-out item with a downward trend, while the chart in the bottom left-hand corner represents a phase-in item with an upward trend. Finally, the chart in the bottom right-hand corner illustrates an item with very low and sporadic demand, which is categorized as an "other" item.

The categorization algorithm involves the following three basic steps:

Step 1: Categorize certain groups of items as "other."

Before beginning the core analysis steps of the algorithm, we first remove items that cannot be categorized into any of the three main categories—steady state, phase-in, or phase-out—from consideration. These include

- items with fewer than eight months of positive demand during the 10-year window of the dataset
- items with fewer than 10 units demanded during the 10-year window of the dataset
- items with fewer than four months of positive demand during the reference year, the two years preceding, and the two years following the reference year
- items with no demand in the reference year.

Step 2: Categorize the remaining items with trivial demand during the reference year.

Once low-demand items are removed, we then categorize items that have trivial demand during the reference year but nontrivial demand before and/or after the reference year. We say that an item has "trivial demand" if the mean item demand during the year is more than one standard deviation below the mean item demand during the 10-year time window of the dataset; otherwise, the item has "nontrivial demand." Note that the definition of *trivial demand* is determined relative to the item's overall demand, not relative to other items or on the basis of absolute demand. For items meeting these criteria, we assign the following labels:

- phase-out—if the item has nontrivial demand prior to the reference year
- phase-in—if the item has nontrivial demand after the reference year
- other—if the item has nontrivial demand both before and after the reference year.

Very few items have trivial demand during the reference year and nontrivial demand both before and after the reference year. Most of these items tend to be low-demand items.

Step 3: Categorize the remaining items using time trend analysis.
To characterize the remaining items, we estimate a linear trend model for each item using demand from the reference year, the previous two years, and the following two years. In total, there are 60 months of demand data used to estimate a time trend. Note that Steps 1 and 2, detailed above, eliminate items with insufficient demand data to apply a statistical approach.

To determine if the data have a trend, we first perform a Kwiatkowski-Phillips-Schmidt-Shin (KPSS) test, which is used to determine whether a time series dataset is stationary.[1] As specified in the algorithm, the null hypothesis of the KPSS test is that the data exhibit a stable trend (i.e., stationary); other specifications of the KPSS test may assess whether the data are stationary with respect to a deterministic trend (i.e., trend stationary). Using a significance level of 95 percent, we categorized items with a p-value above 0.05 as steady-state items. For items for which we rejected the null hypothesis of stationarity, we proceeded to estimate the trend direction, as given by the linear trend model

$$D_t = \beta t + \varepsilon_t,$$

[1] D. Kwiatkowski, P. C. B. Phillips, P. Schmidt, and Y. Shin, "Testing the Null Hypothesis of Stationarity Against the Alternative of a Unit Root," *Journal of Econometrics*, Vol. 54, 1992, pp. 159–178.

where D_t is the number of units demanded in month t, β is the trend, and ε_t is a random term. Items for which β was positive we categorized as phase-in items, while items for which β was negative were categorized as phase-out items.

Algorithm Development

After each test run of the algorithm, we evaluated the results by visually comparing the demand pattern over the 10-year window of the dataset with the assigned categorization for a sample of 80 items (20 per category); we then judged whether the categorization was accurate. From this comparison, we adjusted the rules. This process was repeated until the categorization algorithm achieved 90 percent accuracy in categorizing items based on our judgment.

Specific adjustments were implemented in each step of the algorithm to improve its accuracy. The numbers used in the four categories of items defined in Step 1 were carefully selected to maximize the accuracy of the algorithm. In addition, we evaluated the definition of *trivial demand* in Step 2 of the algorithm to determine if improvements in the performance of the algorithm were possible. Finally, evaluation of the trend estimation in the third step presented several opportunities for improvement. First, we noted that single-month spikes in demand could have outsized effects on the trend direction; in a few cases, removing one data point completely reversed the sign of the trend. To address this issue, we replaced the maximum demand value for each item with the mean demand for that item across all months with positive demand; the time trend model was then reestimated. For instance, if an item had twenty months with an average of 10 units demanded and one month with 100 units demanded (and no demand in any of the remaining months), we replaced the demand value of 100 with 10. Furthermore, our initial results indicated that the 60 months of data used to estimate the trend may lead to an inaccurate or misleading conclusion when considering all 10 years' worth of data. For instance, some items had a negative trend across the 10-year time window, but a positive trend in the 60 months used to estimate the trend in the analysis year. In these cases, we categorized the item as steady state. The algorithm evaluation process was repeated until the categorization algorithm achieved 90 percent accuracy in categorizing items.

While for most items it was obvious whether the assigned category was accurate, for some items, the observed demand pattern did not fit neatly into one of the four major categories. We aimed to avoid obvious categorization errors, recognizing that these "grey" category items will not have a clear correct category. Thus, analysis using the categorization results from this algorithm should be considered generally descriptive of the populations, not a precise characterization of the process performance for each category and supply chain management organization.

The algorithm detailed above is implemented in the *R* programming language, a free statistical software package that runs on Windows, Linux, and Mac operating systems.

APPENDIX F

Illustrating the Consequences of a Push System

Example NIINs to Illustrate the Production of Long Supply

Figures F.1, F.2, and F.3 show three Army NIINs that were repaired into long supply based on the amount of serviceable inventory compared to demand, with many years of serviceable supply having been produced in each case.

Figure F.1
NIIN 014087048 Demand and Inventory History

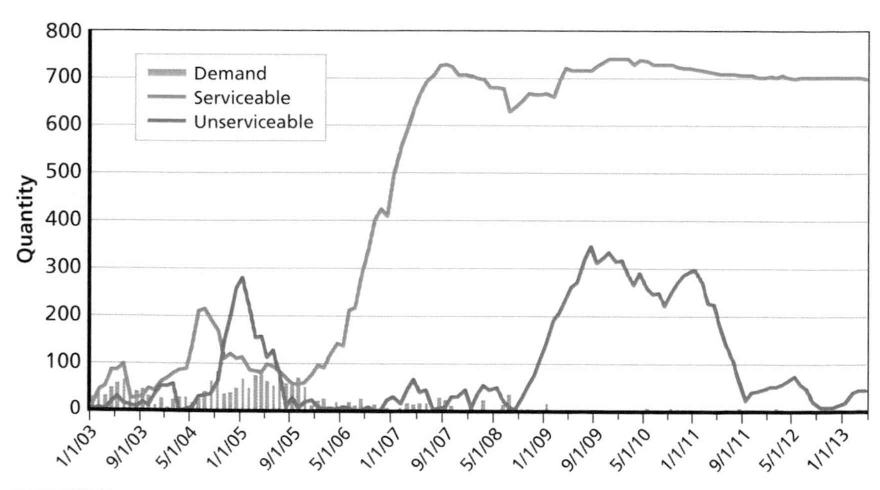

RAND *RR398-F.1*

Figure F.2
NIIN 00592298 Demand and Inventory History

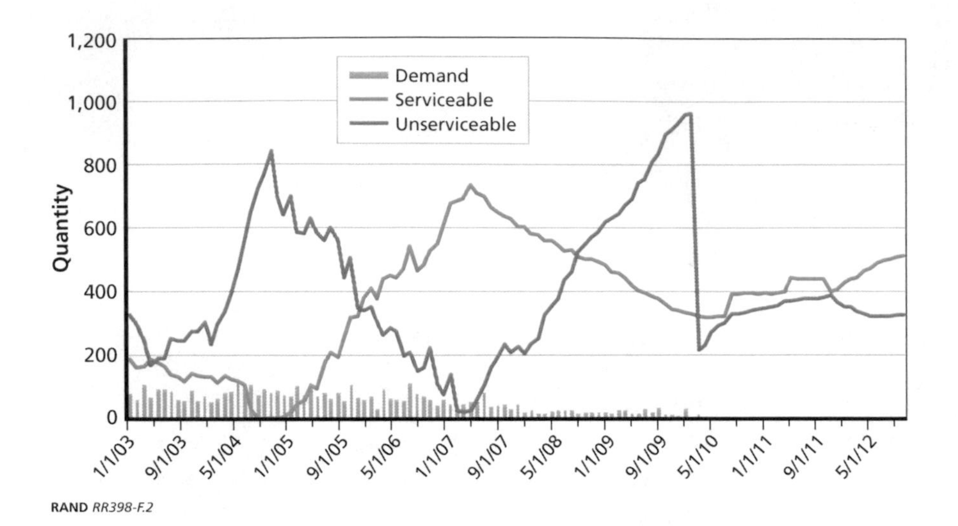

RAND *RR398-F.2*

Figure F.3
NIIN 013332064 Demand and Inventory History

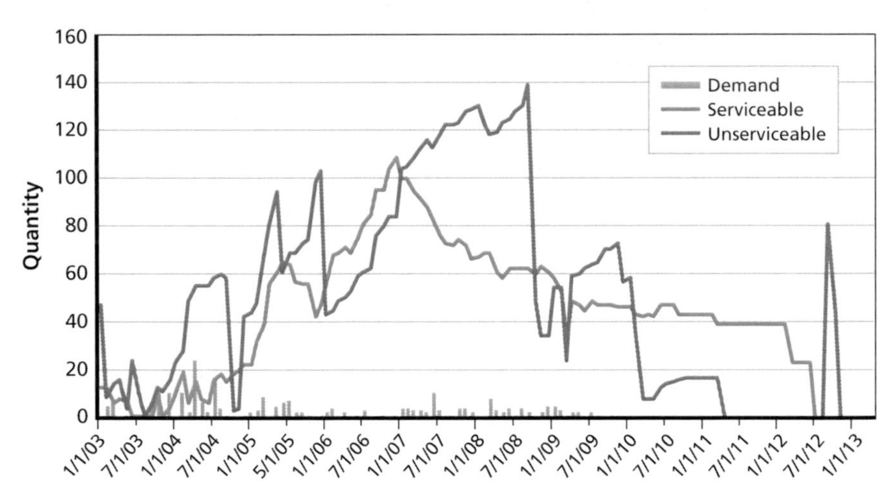

RAND *RR398-F.3*

Demand Versus Repair, Production, and Inventory by Army Inventory Control Point

Figure F.4 shows the value of DLR demands by the LCMC and subordinate inventory control point that manages the part demanded. TACOM-Warren has had a steady demand decline since 2008. Declining demand for AMCOM-air and CECOM began in 2010. The downward demand trend is fairly steady for TACOM-Rock Island (RI) and AMCOM-missile; however, TACOM-RI experienced a small temporary increase in demand in 2010.

After observing the long downward demand trend for AMCOM-missile, we looked to see how repair and buy trends responded and their effect on serviceable inventory. Figure F.5 shows ten years of DLR demands, purchases, repairs, disposals, serviceable inventory, and unserviceable inventory for AMCOM-missile. Demand, shown by the red series, declined in 2004. Repairs (green) and procurements (purple) did not, resulting in an increase in serviceable inventory, but this was necessary to return customer support to appropriate levels of effectiveness. As shown in Figure 4.5, customer support declined dramatically in 2003 across the LCMCs due to insufficient inventory requirements early in OIF and the subsequent depletion of many items with the increase in demand at the start of OIF. This was exacerbated by insufficient funding in FY 2002, which led to a decline in customer support that began in the second half of 2002, even before OIF was launched.[1] As that figure shows, it took a few years for customer support to fully recover as serviceable inventory was built back up. Later, after this recovery, demand again declined from 2007 to 2008, with time repair and procurements declining as well after a lag of a couple of years. This again resulted in an increase in serviceable inventory, potentially more than was needed for some items. Unserviceable inventory increased as well between 2006 and 2010, reflecting an overall increase in inventory until disposals ramped up starting in 2011.

For the AMCOM-air DLR (Figure F.6) D demand began to trend downward after 2010, with the decline in repairs and procure-

[1] Peltz et al., 2005.

Figure F.4
Value of DLR Demands Over Time

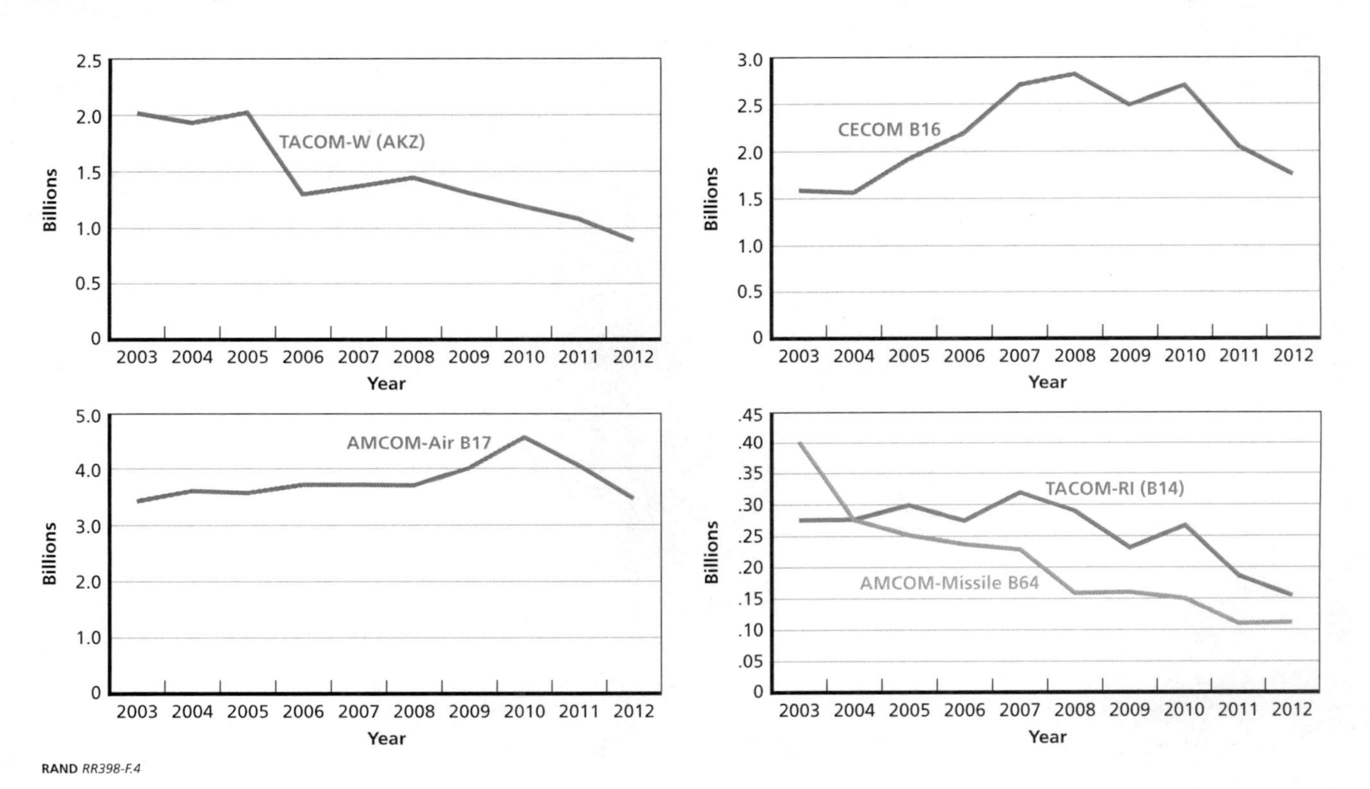

Figure F.5
Ten Years of DLR Demands, Buys, Repairs, and Serviceable Inventory for AMCOM-Missile

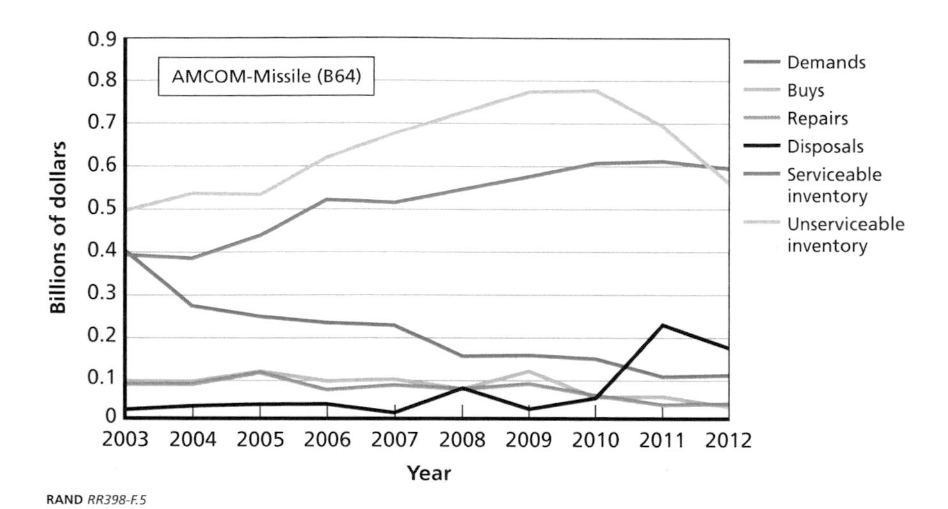

RAND *RR398-F.5*

Figure F.6
Ten Years of DLR Demands, Buys, Repairs, and Serviceable Inventory for AMCOM-Air

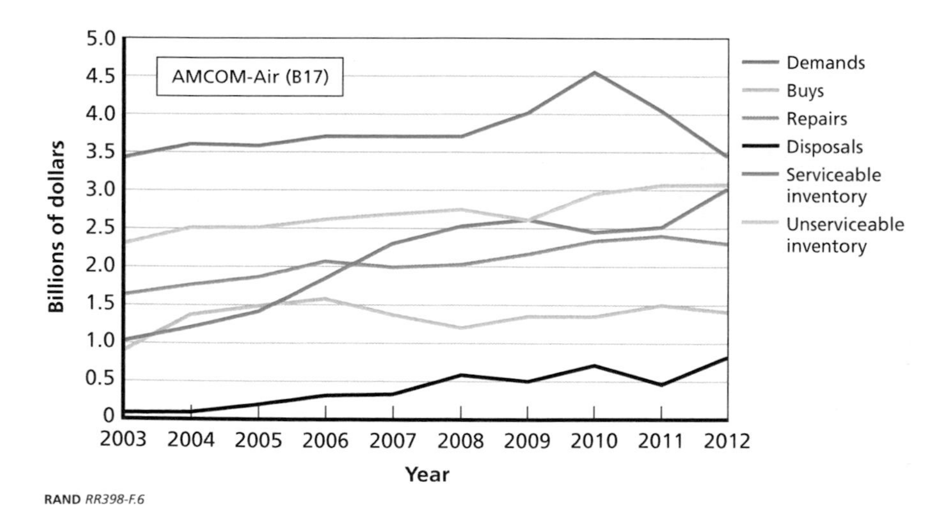

RAND *RR398-F.6*

ments lagging by a year, resulting in some increase in serviceable inventory. Unserviceable inventory and disposals both gradually climbed over the full period.

TACOM-W's DLR trends are graphed in Figure F.7. The demand decrease in 2006 is lagged by the repair and procurement increases, leading to much higher serviceable inventory, although much of this increase was necessary for stock availability to recover. A downward trend starting in 2009 was lagged by a decline in procurements, although repairs responded quickly. The net result, though, was another increase in serviceable inventory. With the decline in demands there was also an uptick in disposals.

The data for TACOM-RI in Figure F.8 shows demand trending downward after 2007. Repairs started declining in 2005, but purchases increased through 2008 and did not decline as steeply as demands. Thus, serviceable inventory grew through 2010, after which it declined. Unserviceable inventory also climbed through 2008, before increasing disposals and declining demand (and repairs) led to decreases.

Figure F.7
Ten Years of DLR Demands, Buys, Repairs, and Serviceable Inventory for TACOM-W

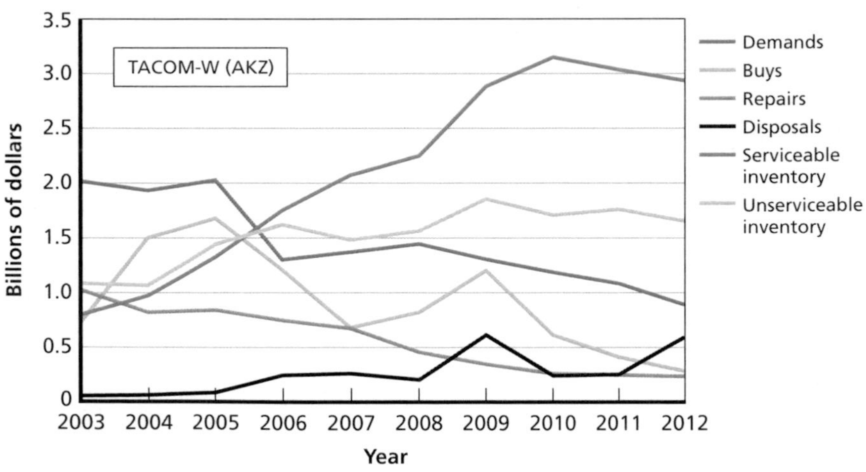

RAND *RR398-F.7*

Figure F.8
Ten Years of DLR Demands, Buys, Repairs, and Serviceable Inventory for TACOM (B14)

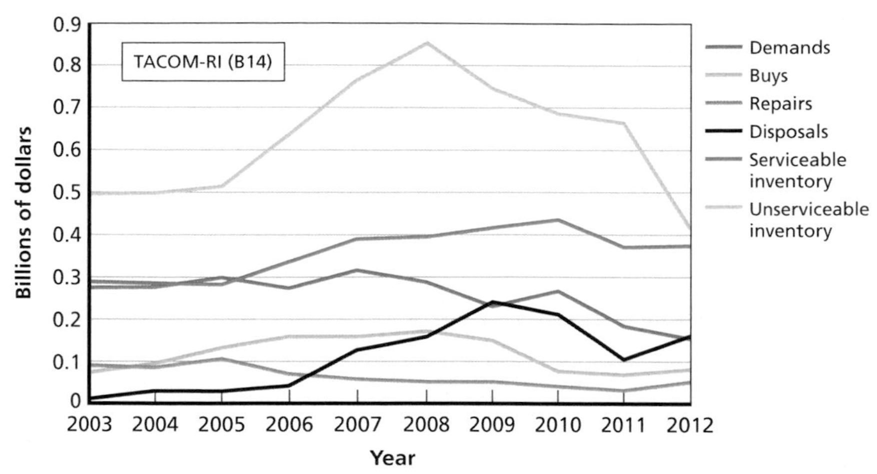

RAND *RR398-F.8*

As shown in Figure F.9, CECOM's demands increased steadily from 2004 to 2008, but purchases increased more rapidly than did demands, leading to a peak in serviceable inventory in 2009. Repairs were steady throughout the ten years. Unserviceable inventory and disposals both gradually climbed over most of the full period.

Figure F.9
Ten Years of DLR Demands, Buys, Repairs, and Serviceable Inventory for CECOM (B16)

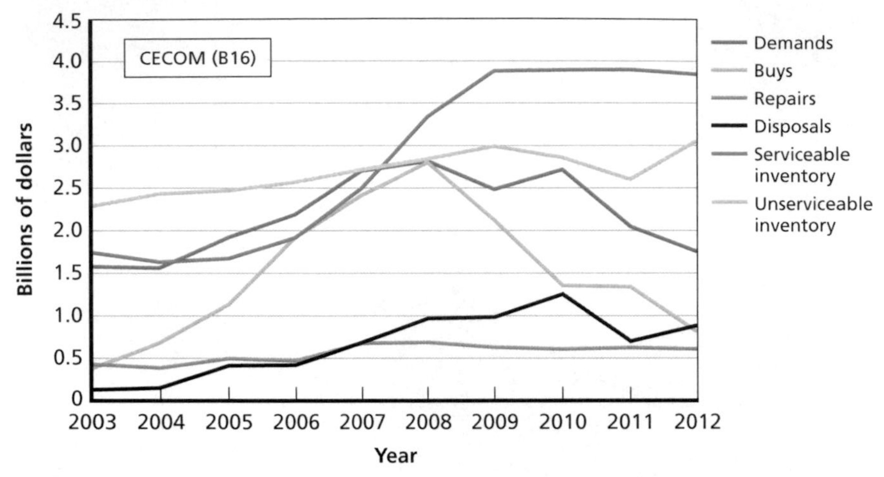

RAND *RR398-F.9*

References

Abell, John B., L. W. Miller, Curtis E. Neumann, and Judith E. Payne, *DRIVE (Distribution and Repair in Variable Environments): Enhancing the Responsiveness of Depot Repair*, Santa Monica, Calif.: RAND Corporation, R-3888-AF, 1992. As of January 10, 2014:
http://www.rand.org/pubs/reports/R3888.html

Airbus, "British Airways Selects Airbus Flight Hour Services for Its A380 Fleet," October 6, 2011. As of November 13, 2013:
http://www.aerospacenewsline.com/articles/2011100600000003.html

Ashayeri, J. R., et al., "Inventory Management of Repairable Service Parts for Personal Computers: A Case Study," *International Journal of Operations & Production Management*, Vol. 16, No. 12, 1996.

Boito, Michael, Cynthia R. Cook, and John C. Graser, *Contractor Logistics Support in the U.S. Air Force*, Santa Monica, Calif.: RAND Corporation, MG-779-AF, 2009. As of July 3, 2013:
http://www.rand.org/pubs/monographs/MG779

Chenoweth, Mary E., Jeremy Arkes, and Nancy Y. Moore, *Best Practices in Developing Proactive Supply Strategies for Air Force Low-Demand Service Parts*, Santa Monica, Calif.: RAND Corporation, MG-858-AF, 2010. As of May 28, 2013:
http://www.rand.org/pubs/monographs/MG858

Cook, Cynthia R., Michael Boito, John C. Graser, Edward G. Keating, Michael J. Neumann, and Ian P. Cook, *A Methodology for Comparing Costs and Benefits of Management Alternatives for F-22 Sustainment*, Santa Monica, Calif.: RAND Corporation, TR-763-AF, 2011. As of July 19, 2013:
http://www.rand.org/pubs/technical_reports/TR763

Creel, Amanda, "Robins First Air Logistics Center to Implement BRAC 2005 Decisions," *Air Force Print News Today*, April 27, 2007. As of March 23, 2013:
http://www.afmc.af.mil/news/story_print.asp?id=123050923

Daly, Kieran, "Cash on the Shelf," *Airline Business*, Vol. 25, No. 11, November 2009.

Day, Allan, "309th Maintenance Wing Overview," Ogden ALC 309th Maintenance Wing, August 16, 2011.

Day, Allan, "Commentary: 309 MXW to Inactivate; Members to Be Part of OO-Air Logistics Complex," *Hilltop Times*, June 7, 2012. As of April 24, 2013: http://www.hilltoptimes.com/content/309-mxw-inactivate-members-be-part-oo-air-logistics-complex

Defense Logistics Agency, "IMSP Spiral 2 Gross Demand Plan Process Flow Overview," briefing, May 15, 2013.

Department of the Air Force, *United States Air Force Working Capital Fund (Appropriation: 4930)*, Fiscal Year (FY) 2013 Budget Estimates, February 2012.

Department of the Army, *Army Working Capital Fund Fiscal Year (FY) 2013 President's Budget*, February 2012.

Department of Defense, *Defense Working Capital Fund, Defense-Wide Fiscal Year (FY) 2013 Budget Estimates Operating and Capital Budgets*, February 2012.

Department of the Navy, *Fiscal Year (FY) 2013 Budget Estimates: Justification of Estimates Navy Working Capital Fund*, February 2012.

Fleischmann, M. J., et al., "Integrating Closed-Loop Supply Chains and Spare-Parts Management at IBM," *Interfaces*, Vol. 33, No. 6, 2003.

Fuentes, Gidget, "End Nears for CH-46E Sea Knight Helicopter," *Marine Corps Times*, August 23, 2008. As of November 13, 2013: http://www.marinecorpstimes.com/article/20080823/NEWS01/808230303/End-nears-CH-46E-Sea-Knight-helicopter

Girz, Tom, "448th Supply Chain Management Wing: Long Term Strategic Plan Update," briefing, January 25, 2011.

"The Global C-17 Sustainment Partnership," *Defense Industry Daily*, January 7, 2013. As of August 16, 2013: http://www.defenseindustrydaily.com/did-focus-the-c17-global-sustainment-partnership-02756/

Headquarters, Air Force and Headquarters, AFMC/LG, "AF/AFMC/DLA Performance Based Agreement Addendum," March 14, 2012.

Headquarters, Air Force Materiel Command, *Material Management: Planning for DLA-Managed Consumables (PDMC)*, Air Force Materiel Command Instruction 23-205, April 26, 2012.

Headquarters, Defense Logistics Agency, Air Force Military Service Support Team, and Headquarters Air Force, ILCM Policy Division, "Logistics, Installations & Mission Support United States Air Force (AF/A4/7) and Defense Logistics Agency (DLA) Performance Based Agreement," Version 3.0, September 20, 2010.

Headquarters, United States Army G-4 and Headquarters, Defense Logistics Agency, Performance Based Agreement, May 12, 2008.

Hill Air Force Base, "U.S. Air Force Fact Sheet: AFLC," undated. As of March 28, 2013:
http://www.hill.af.mil/library/factsheets/factsheet_print.asp?fsID=5594

Keating, Edward G., Adam C. Resnick, Elvira N. Loredo, and Richard Hillestad, *Insights on Aircraft Programmed Depot Maintenance: An Analysis of F-15 PDM*, Santa Monica, Calif.: RAND Corporation, TR-528-AF, 2008.

Krikke, Harold, and Erwin van der Laan, "Last Time Buy and Control Policies with Phase-Out Returns: A Case Study in Plant Control Systems," *International Journal of Production Research*, Vol. 49, 2011.

Kwiatkowski, D., P. C. B. Phillips, P. Schmidt, and Y. Shin, "Testing the Null Hypothesis of Stationarity Against the Alternative of a Unit Root," *Journal of Econometrics*, Vol. 54, 1992.

LaFalce, John T., "AMC Repair Parts Supply Chain," *Army Logistician*, May–June 2009. As of April 2013:
http://www.almc.army.mil/alog/issues/may-june09/AMC_repchain.html

Marine Corps Team, Military Service Support Division, and Defense Logistics Agency, Performance Based Agreement Between the Defense Logistics Agency and Headquarters, U.S. Marine Corps, Version 3.0, July 8, 2010.

Mecham, M., "Golden TUI," *Aviation Week and Space Technology*, Vol. 172, No. 15, 2010.

Miller, L. W., and John B. Abell, *DRIVE (Distribution and Repair in Variable Environments): Design and Operation of the Ogden Prototype*, Santa Monica, Calif.: RAND Corporation, R-4158-AF, 1992. As of January 10, 2014:
http://www.rand.org/pubs/reports/R4158.html

Moorman, Robert W., "OEMs Tout Savings Via Rotable Programs," *Aviation Week*, Vol. 17, No. 1, January 2011.

Naval Supply Systems Command, "Navy Retrograde to Army G-4," briefing, January 27, 2011.

Office of Management and Budget, *Guidelines and Discount Rates for Benefit-Cost Analysis of Federal Programs*, Circular No. A-94, October 29, 1992 (Appendix C, Revised December 2011).

Ogden Air Logistics Center Public Affairs Office, "Ogden Air Logistics Center," undated. As of March 28, 2013:
http://www.hill.af.mil/shared/media/document/AFD-100409-048.pdf

Peltz, Eric, and Marc Robbins with Geoffrey McGovern, *Integrating the Department of Defense Supply Chain*, Santa Monica, Calif.: RAND Corporation, TR-1274-OSD, 2012. As of January 10, 2014:
http://www.rand.org/pubs/technical_reports/TR1274.html

Peltz, Eric, Marc Robbins, Kenneth Girardini, Rick Eden, and Jeffrey Angers, *Sustainment of Army Forces in Operation Iraqi Freedom: Major Findings and Recommendations*, Santa Monica, Calif.: RAND Corporation, MG-342-A, 2005. As of January 10, 2014:
http://www.rand.org/pubs/monographs/MG342.html

Performance Based Agreement Between Director, Defense Logistics Agency and Chief of Naval Operations and Commander, Naval Supply Systems Command signed February 16, 2011, March 25, 2011, and March 8, 2011, respectively.

Ray, Mike W., "Alsup Named 448th SCMW Director," *Air Force Print News Today*, March 15, 2013. As of April 29, 2013:
http://www.tinker.af.mil/news/story_print.asp?id=123340323

Richardson, H. L., "Service Parts: Reaching the Right Level," *Transportation & Distribution*, Vol. 39, No. 9, 1998.

Robins Air Force Base, "Warner Robins Air Logistics Complex," undated. As of March 28, 2013:
http://www.robins.af.mil/units/wrairlogisticscomplex.asp

Seidenman, Paul, and David Spanovich, "Reconfiguring Avionics Support," *Aviation Week Overhaul and Maintenance*, Vol. 18, No. 5, May 2012.

Simpson, Susan, "Tinker Employees Fill Former GM Plant," NewsOK.com, August 22, 2010. As of November 22, 2013:
http://newsok.com/tinker-employees-fill-former-gm-plant/article/3487516

Tedone, Mark J., "Repairable Part Management," *Interfaces*, Vol. 19, No. 4, July–August 2009.

Tinker Public Affairs, "U.S. Air Force Fact Sheet: Oklahoma City Air Logistics Complex," August 28, 2012. As of March 28, 2013:
http://www.tinker.af.mil/library/factsheets/factsheet.asp?id=8552

Tripp, Robert S., Kristin F. Lynch, Daniel M. Romano, William L. Shelton, John A. Ausink, Chelsea Kaihoi Duran, Robert G. DeFeo, David W. George, Raymond E. Conley, Bernard Fox, and Jerry M. Sollinger, *Air Force Materiel Command Reorganization Analysis: Final Report*, Santa Monica, Calif.: RAND Corporation, MG-1219-AF, 2012. As of January 10, 2014:
http://www.rand.org/pubs/monographs/MG1219.html

United States Code, Title 10, Section 2464, Core Depot-Level Maintenance and Repair Capabilities. As of August 17, 2013:
http://www.law.cornell.edu/uscode/text/10/2464

United States Code, Title 10, Section 2466, Limitations on the Performance of Depot-Level Maintenance of Materiel. As of August 17, 2013:
http://www.law.cornell.edu/uscode/text/10/2466

U.S. Air Force, "Air Force Materiel Command," January 30, 2013.

U.S. Air Force, Air Force Materiel Command Instruction 23-120, *Material Management: Execution and Prioritization Repair Support System (EXPRESS)*, May 24, 2006.

U.S. Air Force Materiel Command Public Affairs, "AFMC Restructures to Cut Overhead, Make Command More Efficient," November 18, 2011. As of April 21, 2013:
http://www.afmc.af.mil/news/story.asp?id=123278315

U.S. Air Force Sustainment Center, "Air Force Sustainment Center," July 20, 2012. As of April 2, 2013:
http://www.afsc.af.mil/library/factsheets/factsheet.asp?id=19642

U.S. Army, *Organic Industrial Base Strategic Plan (AOIBSP) 2012–2022*, undated. As of April 19, 2013:
http://usarmy.vo.llnwd.net/e2/c/downloads/276549.pdf

U.S. Army Materiel Command, *National Maintenance Program Business Process Manual*, Chapter 4: National Workload Plan Development and Validation, October 1, 2012.

U.S. Government Accountability Office, "Navy Regional Maintenance: Substantial Opportunities Exist to Build on Infrastructure Streamlining Progress," 1997. As of November 13, 2013:
http://www.gao.gov/assets/230/224909.pdf

U.S. Government Accountability Office, *Oversight and a Coordinated Strategy Needed to Implement the Army Workload and Performance System,* GAO-11-566R, July 14, 2011.

U.S. Government Accountability Office, "Military Base Realignments and Closures: Updated Costs and Savings Estimates from BRAC 2005," GAO-12-709R, June 29, 2012. As of March 31, 2013:
http://www.gao.gov/assets/600/592076.pdf